Exercises for the BOTANY LABORATORY

Joel A. Kazmierski
Professor Emeritus
State University of New York
College of Technology at Delhi

Morton Publishing Company
925 W. Kenyon Ave., Unit 12
Englewood, Colorado 80110
http://www.morton-pub.com

Book Team

Publisher	Douglas Morton
Copy Editor	Kevin Campbell, Greenleaf Editorial
Production Manager	Joanne Saliger, Ash Street Typecrafters, Inc.
Cover Design	Bob Schram, Bookends
Design & Typography	Ash Street Typecrafters, Inc.

Printed in the United States of America
by Morton Publishing Company, 925 W. Kenyon Ave., Unit 12, Englewood, CO 80110

10 9 8 7

ISBN: 0-89582-489-2

ISBN-13: 978-0-89582-489-9

Contents

Student Information

Name ______________________________________

School Address ________________________________

Telephone Number ____________________

Lab Day and Time ____________ / ______________

Instructor ____________________________________

Office ______________________________________

Telephone ____________________

Preface

The laboratory probably is the most important learning center for any biology course. This is where the student actually gets to see, touch, and experience much of the information explained in the lecture portion of the course. However, it still remains the student's responsibility to make maximum use of this time to learn as much as possible.

This laboratory manual is designed to introduce the beginning botany student to the basic features of plant structure and diversity. The first seven exercises deal with the morphology and anatomy of seed plants, especially the flowering plants. The remaining exercises are given to a survey of the plant kingdom. In the latter exercises, some of the basic concepts of plant evolution will be examined, as well as the morphological and anatomical features of representative plants.

I would like to make the following suggestions for use of this manual:

1. Bring your text to lab. The additional information and good illustrations in it will prove very useful in understanding certain portions of many exercises.
2. Use the blank pages preceding each exercise for taking introductory notes.
3. A list of goals or objectives is given at the beginning of each exercise. Work on a particular unit should not be considered complete unless each of these goals has been met.
4. Each exercise has photographs of many of the items to be studied. All of these should be *labeled as completely as possible*. It should be emphasized, however, that these illustrations are not meant to replace the observation of the actual specimen.
5. Most exercises also require some diagrams of specified material. These diagrams should be large enough to show sufficient detail by using most of the allotted space, and they should be as accurate as possible (not necessarily a work of art), done in *pencil,* and should be well labeled. Good diagrams are extremely useful learning aids because (1) they require detailed observation of the material, (2) the transmission of this observed information into a diagram will help in remembering the material, and (3) the diagrams will be good study guides. The following figures illustrate the difference between good and poor diagrams.

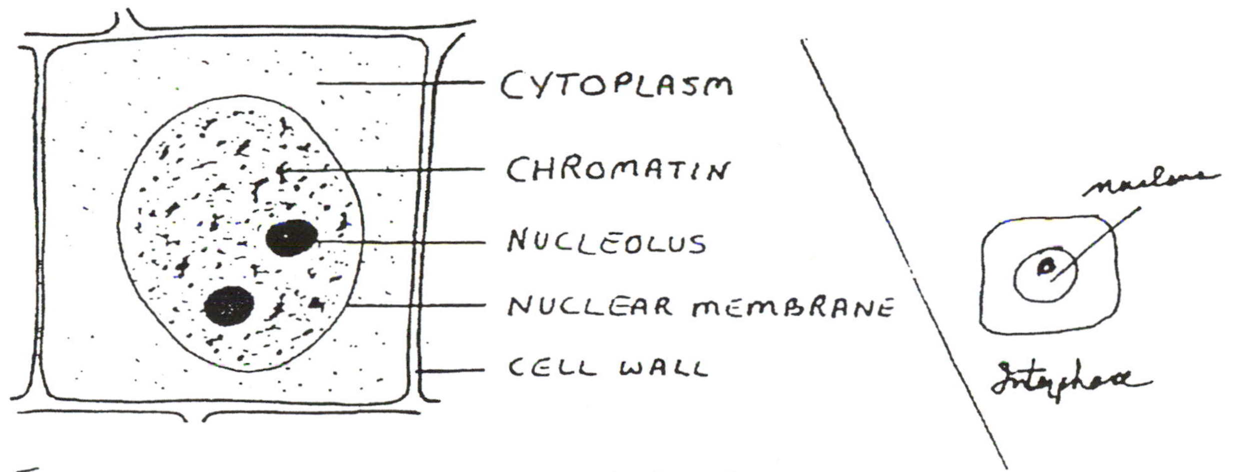

Fig. 1 Onion Root Tip Cell in Interphase

6. Throughout the exercises, and at the end of each one, are a number of questions. All of these questions should be answered, ideally by using information obtained by doing the exercise. Discussing the more difficult questions with other members of the class is a useful procedure, much better than just asking the instructor for an answer. If you find that you still cannot come to a conclusion, however, be sure to ask the instructor.
7. Finally, be sure to *do* all the work described in each unit. Unless told otherwise, follow the order presented in the exercise. Merely labeling the diagrams and answering the questions does not fulfill your obligations to the course. For the most part, you should make your own observations and do your own work. There are relatively few exercises that call for teamwork.

Cover Information

The cover is a photograph of the camas lily (*Camassia quamash*). In the late springtime, it produces very pretty blue flowers in clusters at the end of a stalk that may be more than two feet tall. It often grows in large patches in moist meadows of the intermountain region of Idaho and Montana (it may be found as far south as California). The plant may be so abundant as to give the appearance of a small lake or pond on the middle of the meadow (exactly what my two-year-old son thought in 1971). In 1806, Meriwether Lewis described the plant in a very similar manner.

> The quawmash [camas] is now in blume and from the color of its bloom at a short distance it resembles a lake of fine clear water, so complete is this description that on first sight I could have swoarn it was water.[1]

The bulb of this plant was once a very important food source for the Native Americans of that region. At times, members of the Lewis and Clark expedition were entirely dependent on it for food, although their writings about its flavor were never very favorable. Consider the following comments, the first by Meriwether Lewis, the second by William Clark:

> The root [bulb] is pallateable but disagrees with me in every shape I have ever used it.[2] (*Shape*, refers to the various methods of preparing the food.)

> Capt Lewis & my Self eate a Supper of roots boiled, which filled us So full of wind that we were Scercely able to Breathe all night felt the effects of it.[3]

Father Nicholas Point, an early missionary to the region, provided much information about camas; its appearance, how it was gathered by the women and children, the tools they used, how it was cooked, and how it was preserved. He also added the following commentary:

> The root tastes somewhat like a prune and a chestnut. It is eaten with pleasure, but its digestion is accompanied by very disagreeable effects for those who do not like strong odors or the sounds that accompany them.[4]

One of the last armed confrontations between Native Americans and U.S. soldiers came about, in part, because of the native's dependence on camas for food. The peaceful Nez Percé tribe, which occupied the fertile lands at the Idaho, Washington, and Oregon junction, was moved to a reservation along the Clearwater River in Idaho, a land which is very rocky and not suitable for the growth of large quantities of camas. The tribe left the reservation to collect the bulbs, an infraction of the restrictions the government had placed on them. This was the start of the Chief Joseph War.

If you ever want to read a moving and heroic account of how a small band of people desperately tried to defend their homeland and preserve their culture against overwhelming odds, read an account of this conflict. In some four months, Chief Joseph led his people on a trek of some 1,300 miles across Montana toward Canada. Despite all attempts by the tribe to avoid conflict, there were several skirmishes, but only the last one was a clear victory for the soldiers. Here is a portion of Chief Joseph's surrender speech:

> Tell General Howard I know his heart. What he told me before I have in my heart. I am tired of fighting. Our chiefs are killed. Looking Glass is dead. Toohoolhoolzote is dead. The old men are dead. It is the young who say yes or no. The little children are freezing to death. It is cold and we have no blankets. My people, some of them, have run away to the hills, and have no blankets; no one knows where they are — perhaps freezing to death. I want to have time to look for my children and see how many I can find. Maybe I shall find them among the dead. Hear me my chiefs. I am tired; my heart is sick and sad. From where the sun now stands, I will fight no more forever.[5]

Chief Joseph was not given the time he hoped for — time to look for his children. Instead, he was moved to a reservation in the desert country of Washington, a region much different from the land that he had tried to defend. He died there, never having been allowed to leave and return to his native homeland.

Dedication

This is dedicated to two ladies who greatly stimulated my curiosity about things in the natural world. My mother, Louella Kazmierski Gooley, has always been keenly interested in nature, but is especially interested in wildflowers and other plants. She still asks me questions about things she has observed around her house in Ashby, Massachusetts. I think I sometimes disappoint her by admitting that I don't know the answer to the question. When I went to college, I came under the influence of Dr. Helen Ross Russell, a former student of E. Lawrence Palmer at Cornell University. I have fond memories of wading in streams or ponds with her as she and the class enthusiastically seeked out the denizens of these habitats. It was through her that I received my first lessons in plant systematics as we took field trips through fields and woods.

1. As quoted in Ambrose, Stephen E. 1996. *Undaunted Courage: Meriwether Lewis, Thomas Jefferson, and the opening of the American west.* New York: Simon & Schuster.
2. Ibid.
3. As quoted in Dayton, Butron, & Ken Burns. 1997. *Lewis & Clark: The journey of the Corps of Discovery. An illustrated history*. New York: Alfred A. Knopf.
4. As quoted in Johnston, A. 1970. Blackfoot indian utilization of the flora of the northwestern Great Plains. *Economic Botany 24* (3), 301–24.
5. Joseph A., Jr. 1964. *Chief Joseph's people and their war*. Bozeman, MT: Yellowstone Library and Museum Association.

Acknowledgments

The photographs of onion root tip cells in various stages of mitosis (Figs. 2.2–2.8, 2.13) and the asexual stage of *Rhizopus* (Fig. 8.5) were copied from 35 mm transparencies with the permission of Carolina Biological Supply Company, Burlington, North Carolina.

The photograph of a longitudinal section of a corn kernel (Fig. 15.6) was copied from a 35 mm transparency, courtesy of Turtox/Cambosco, Chicago, Illinois.

The photograph of a compound microscope (Fig. 1.1) is courtesy of Leica Microsystems.

All other photographs and diagrams are by the author.

Microscopes and Plant Cell Structure

Introductory Notes

The microscope is undoubtedly the most important tool used in the study of general botany. With proper use, objects much too small to be seen with the unaided eye can be seen in great detail. Since you will be using microscopes frequently, it is important that you understand their capabilities, limitations, how they work, and how to care for them.

THE DISSECTING MICROSCOPE

The dissecting microscope has many uses and is extremely simple to operate. Your instructor will explain how to use it, after which you should take a few minutes to observe the objects provided.

Give close attention to the image seen through this microscope, so you can compare it with that seen through a compound microscope.

GOALS

After completing this exercise, the student should be able to:

- explain the advantages and limitations of both the dissecting microscope and the compound microscope.
- recognize and identify the main components of a compound microscope.
- explain the functions of the main components of a compound microscope.
- describe several procedures that involve the proper care of a microscope.
- demonstrate and/or explain how to make a wet mount.
- demonstrate how to set up a compound microscope correctly and also locate and focus on an object within a reasonable length of time (one minute).
- compute the total magnification of a compound microscope when given the magnifying power of the various lenses.
- make a labeled diagram of a typical plant cell and be able to recognize the main cellular organelles from a photograph or diagram.
- explain the differences between chloroplasts, chromoplasts, and leucoplasts.
- answer the questions in the exercise.

THE COMPOUND MICROSCOPE

Parts of the Microscope

Obtain a microscope from the cabinet. The microscopes are numbered, and you should take the one that corresponds to your seat number. This is the microscope you should use throughout the rest of the semester. Be sure to carry it with both hands, upright, and take care not to bump it against any of the lab table tops.

Do not attempt to use the microscope until your instructor has discussed the instrument with you. The main parts of the instrument will be pointed out, and the function of each part will be explained.

Find each of the following parts on your microscope as each is discussed. Each of these should be labeled in Figure 1.1.

1. The heavy, U-shaped support of the microscope is the **base**.
2. The **arm** is the sturdy, upright portion that supports the prism housing and lenses of the instrument.
3. The cylindrical or boxlike portion at the end of the arm is the **body** or **prism housing**. This contains prisms that deflect the light rays into the eyepieces.
4. The **revolving nosepiece** is the movable structure at the lower part of the prism housing to which the objective lenses are attached.
5. The four lens systems attached to the revolving nosepiece comprise the **objective lenses**. Information about these lenses follows.

Objective	Color Code	Objective Magnification
Scanning	blue	4X
Low power	green	10X
High power	yellow	40X
Oil immersion	red	100X

6. The **oculars,** or **eyepieces**, are the lens systems at the top of the prism housing. The ocular magnifies the image produced by the objective lens. Our oculars magnify this image another ten times (10X).

7. The platform on which the glass holding the material to be studied is placed is the **stage.**
8. The mechanical stage is the apparatus used to move the slide around on the stage to observe different parts of the specimen. There are three parts: (1) the **mechanical stage** itself, with reference numbers, (2) a pair of **mechanical stage adjustment knobs** under the stage (the lower knob moves the mechanical stage left and right, and the upper knob moves the apparatus forward and backward), (3) an angled **lever** at the side and on the surface of the stage for holding the slide in place. **Note:** *This lever does not go on top of the slide, but along on edge of it.*
9. Our microscopes have a built-in **light source.** If the microscope is not to be used for several minutes, the light should be turned off.
10. The lens seen beneath the opening in the stage is the **substage condenser.** This focuses a cone of light on the specimen. For most work, this lens should be at its upmost position.

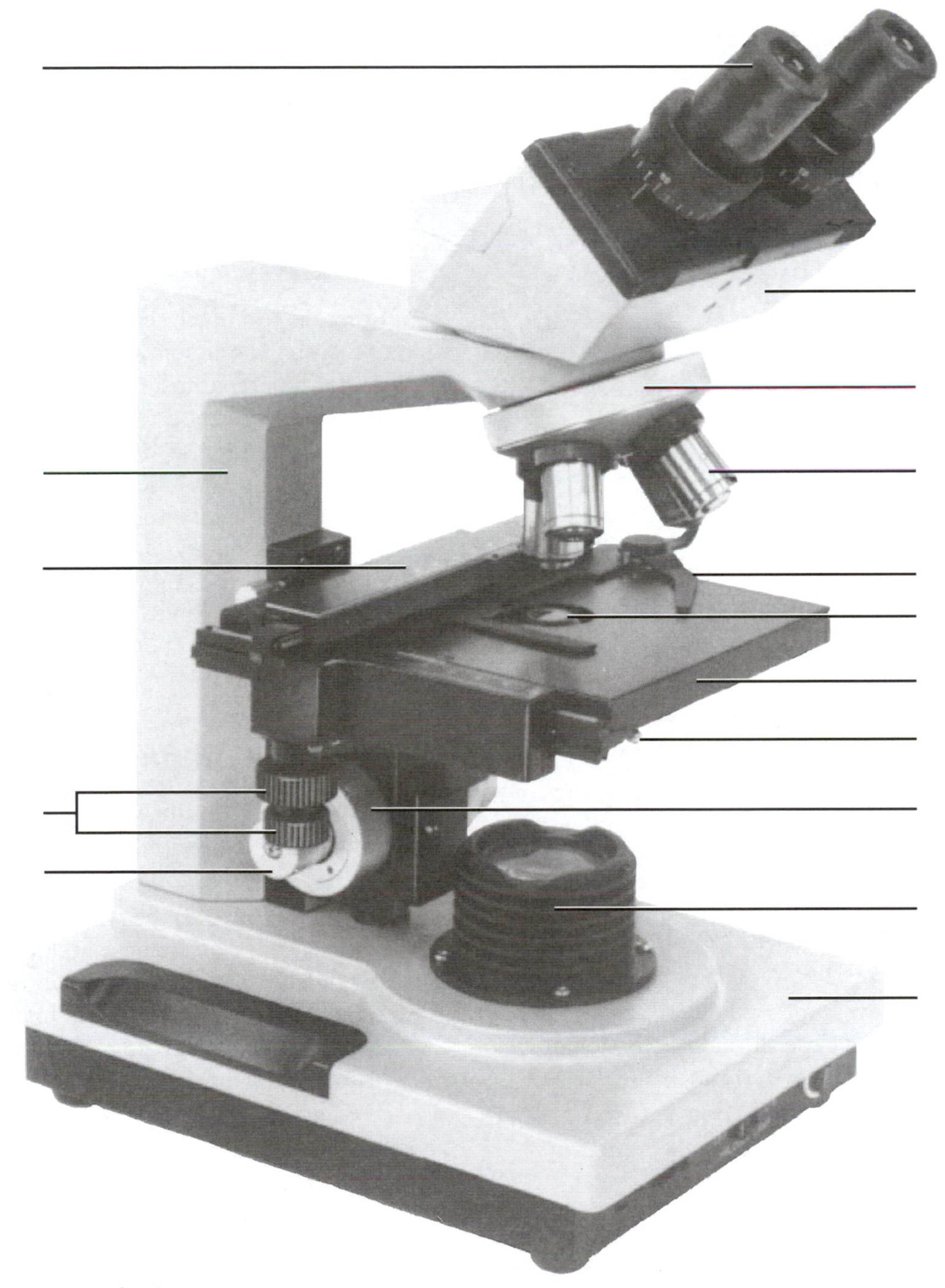

FIGURE 1.1 A compound microscope.

11. The **condenser adjustment knob** is the black knob under the stage (on the side opposite the mechanical stage knobs). It raises or lowers the substage condenser. (This is not visible in the photograph.)
12. The **iris diaphragm** is that part of the substage mechanism used to regulate the light intensity. There is a lever under the stage that is used to adjust the size of the diaphragm aperture. When the low power or scanning objective is in use, the diaphragm opening should be fairly small.
13. The **coarse adjustment knob** is the innermost and larger of the two silver knobs at the base of the stage complex. This is used to raise or lower the stage, bringing the specimen into approximate focus.
14. The smaller of the two focusing knobs is the **fine adjustment knob**. It is used for more precise focusing.

Use and Care of the Microscope

1. If you experience *any* mechanical or optical difficulty, notify your instructor at once.
2. Keep the microscope clean at all times. Use *only lens tissue to clean the lenses.* Never use paper toweling or cloth; these may scratch the glass. If you have difficulty cleaning a lens, ask your instructor for assistance.
3. Do not let the objectives touch any water. If this happens, a distorted image will result. If the lens is not cleaned after this, ultimately, a thin deposit will form on the lens surface, adversely affecting resolution.
4. Always lower the stage slightly and have the low-power objective in place when placing a slide on the microscope or when removing a slide. *Every observation should start with low power.*
5. When using high power, do not focus with the coarse adjustment knob.
6. Try to keep both eyes open when using the microscope. This may be a little difficult at first, but this practice will help reduce eye strain.
7. Always carry the microscope with two hands, upright, and by the arm. If it is tipped too far, the ocular may fall out.
8. Always store the microscope in its proper place, cleaned, and with the cord neatly wrapped around it.

Wet Mounts

Wet mounts are usually used to examine living materials and other materials for which a permanent preparation is not needed. These are very easy to make. Just place a drop of water in the center of a clean microscope slide. Then put a small amount of the material to be examined in the water. Place one edge of a cover slip beside the drop and lower the rest of the cover slip onto the water and over the specimen (Fig. 1.2). Don't just drop the cover slip straight down, because this will result in many air bubbles being trapped under the cover slip.

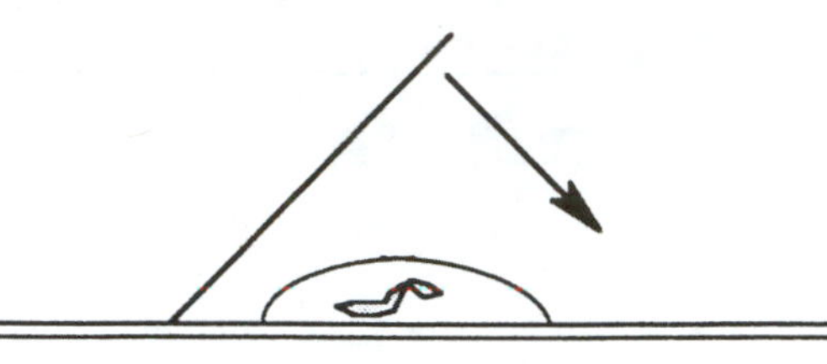

FIGURE 1.2 Wet mount preparation.

If there is too much water, blot up the excess with a paper towel. (Too much water will make the cover slip float, and the preparation will move about and will not be flattened for good observation.) If there is not enough water, simply add some at the edge of the cover slip, and the water will flow underneath. Do this also if the preparation begins to dry out. *The specimen being observed should always be surrounded by water.* Excess air bubbles can be removed by tapping gently on the cover slip. Finally, don't let the bottom of the slide get wet. This will restrict the movement of the slide on the microscope stage.

Use of the Compound Microscope

1. Make a wet mount of a lower case letter *e* from the newspaper (choose one about this size — e). Orient the letter *e* in its normal reading position.
2. Set up the microscope with the arm facing you and with the edge of the base about two or three inches away from the edge of the lab table. Plug the cord into an electrical outlet, and make sure that the excess cord is not hanging over the edge of the table. Adjust your stool to a comfortable height. Turn the microscope on.
3. Be sure the low power (10x) objective is in place. Place the letter *e* slide into the mechanical stage apparatus.
4. Center the letter over the substage condenser lens by turning the mechanical stage adjustment knobs.

5. Raise the stage as far as possible, using the coarse adjustment knob.
6. While looking through the ocular, turn the coarse adjustment knob in a manner that makes the stage move down. Watch for the letter or the paper to come into focus and, when it does, stop.
7. If necessary, position the slide to see the entire letter *e*. Bring the letter into sharp focus by turning the fine adjustment knob.
8. Compare the letter *e* as it is mounted on the slide with the image seen through the microscope. What are two significant differences?

   Orientation: (1)

   Other: (2)

9. Using the mechanical stage adjustment knob, move the slide from left to right. Which way does the image move? (3)
10. If the slide is moved away from you, which way does the image move? (4)
11. Now examine the letter *e* with high power.
    a. Place a printed part of the letter in the center of the field of view.
    b. Turn the revolving nosepiece, and click the high-power objective in place. These microscopes are **parfocal**, which means that little or no refocusing should be necessary when changing magnifications. If the image is slightly out of focus, *use only the fine adjustment knob to correct this.*
    c. Center the portion of the letter you now see.
12. What happened to the *light intensity* when you changed from low power to high power? (5)
13. Is more or less of the letter visible with high power compared to what was seen with the low-power objective in place? (6)
14. Which has the greater depth of field, high power or low power? (7)
15. The circles in Figures 1.3 and 1.4 represent the limits of the microscope field of view. Diagram the letter *e* as seen with low power and high power. Try to make reasonably accurate diagrams.

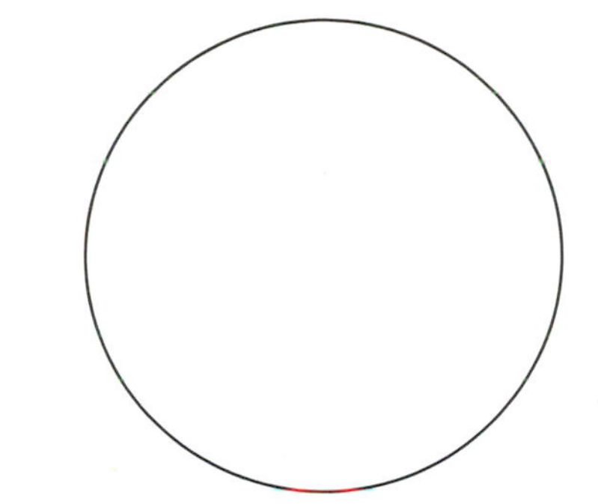

FIGURE 1.3 Letter *e* as seen when using low power.

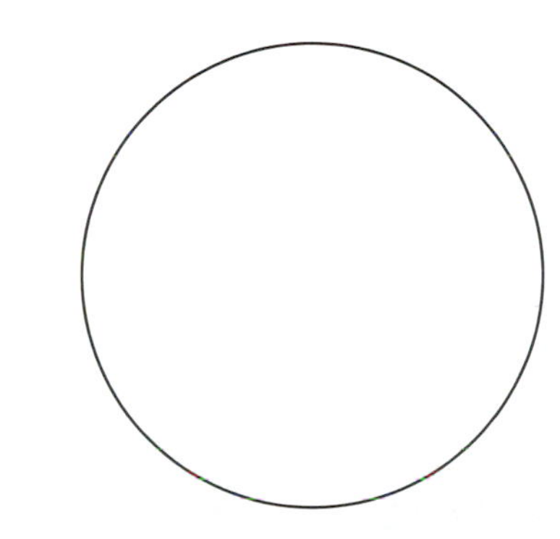

FIGURE 1.4 Letter *e* as seen when using high power.

16. Repeat steps 8–10 using the dissecting microscope. Answer question 4 on page 10.

Determining the Total Magnification

A compound microscope has two lens systems: the objective lenses and the ocular lenses. The objective lens first produces a magnified image of the specimen, and the ocular lens then magnifies this image. The total magnification is the product of these two events. For example:

Objective	×	Ocular	=	Total Magnification
10X	×	10X	=	100X
100X	×	15X	=	1500X

What is the **total magnification** of your microscope for each objective?

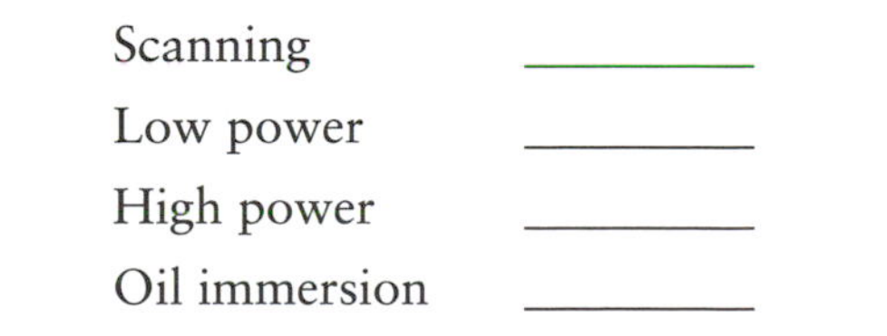

Scanning ________ (8)
Low power ________
High power ________
Oil immersion ________

PLANT CELL STRUCTURE

The cell is the basic structural unit of all organisms. Plant cells have several features that make them different from the cells of animals. In this exercise, you will examine several cells that exemplify these features. Keep in mind, however, that there are many kinds and variations of plant cells. Not all plant cells will resemble the ones you will see today.

Cork Cells

The existence of cells was first noted in 1665 when Robert Hooke described the cellular nature of cork tissue. For this he is given credit for discovering cells. This exercise is designed for you to duplicate his observations.

1. Prepare a dry mount of a small, *very thin* section of cork tissue. (The section does not have to be much larger than the period at the end of this sentence.) Do not use water or add a cover slip.
2. Observe this tissue with both low and then high power.
3. Notice the fairly uniform size and general appearance of these cells. These are dead cells with no internal contents, just a thick, waxy cell wall.
4. Diagram a group of five connected cork cells in the following space, and label the **cell wall** (Fig. 1.5).

Figure 1.5 Cork cells.

Onion (*Allium cepa*) Epidermal Cells

1. Make a wet mount of a small portion of the epidermis from the *inside part* of an onion bulb scale. Be sure the onion tissue is spread out flat.
2. Examine the preparation with low power, and find the region that seems best for more detailed observations. Switch to high power.
3. Try to locate the following structures:
   a. The **cell wall**. What is its function? (9)

   b. The **nucleus** will appear as a spherical structure in the cytoplasm. One or more tiny **nucleoli** (sing. = **nucleolus**) should be visible inside the nucleus.

   c. The **cytoplasm** will appear as a nearly transparent, slightly granular, substance just inside the cell wall, and surrounding the nucleus.

   Clear areas in the cell represent fluid-filled cavities, the **central vacuoles** — another distinct plant cell feature.

   d. If you have difficulty finding these structures, add some IKI (iodine-potassium iodide) solution to your preparation. This will act as a stain and will make the nucleus and cytoplasm more visible.
4. Diagram a single epidermal cell in the following space, and label the **cell wall, cytoplasm, central vacuole, nucleus,** and **nucleolus** (Fig. 1.6).

FIGURE 1.6 Onion epidermal cell.

The next three exercises will introduce you to cellular structures known as **plastids**. These structures are involved in food production or food storage, and there are three major types: chloroplasts, chromoplasts and leucoplasts.

Elodea Leaf Cells

1. Make a wet mount of an entire *Elodea* leaf.
2. Examine it with low power, and select an area midway between the edge of the leaf and the midrib. Switch to high power.
3. Using fine adjustment, focus up and down to determine how many cell layers thick the leaf is. How many layers are present? (10)
4. Find the following cell parts:
   a. The **cell wall.**
   b. The **cytoplasm** and **nucleus** of these cells are very difficult to see and require careful adjustment of the light intensity.
   c. The numerous, oval, green structures in the cytoplasm are **chloroplasts.** In some cells, the chloroplasts may be seen moving around the periphery of the cell. This is a result of a movement of the cytoplasm (cytoplasmic streaming or **cyclosis**).
   d. The **central vacuole** is a clear, water-filled region in the central region of the cell. You won't be able to see the boundary of the vacuole because the vacuolar membrane, or **tonoplast,** cannot be resolved with the light microscope. At best you can see that the chloroplasts and cytoplasm are distributed near the cell wall — nothing is in the middle.
5. Make a diagram of a single *Elodea* cell in the following space (Fig. 1.7). Label the **cell wall, chloroplasts, central vacuole,** and **cytoplasm.**

FIGURE 1.7 *Elodea* cell.

Potato Tuber Cells (*Solanum tuberosum*)

1. Cut a small, *very thin* section from a fresh potato, and make a wet mount. A piece about the size of this letter *o* is adequate.
2. Examine this with low power, and find an area at the edge of the section where intact cells are visible. What is the general shape of these cells? (11)

   Are the cell walls relatively thick or thin? (12)
3. The numerous, irregularly shaped ovoid structures are **leucoplasts.** Several types of leucoplasts are present in plants, all of which are involved with storage of complex organic molecules (food storage). These particular leucoplasts store starch and are known as **amyloplasts.**
4. Notice that these cells do not fit together as tightly as the epidermal cells or cork cells and that there are small triangular **intercellular spaces** (air spaces where three or more cells meet).

5. Diagram three potato cells in the following space (Fig. 1.8). Label the **cell wall, amyloplasts,** and **intercellular space.**

FIGURE 1.8 Potato tuber cells with amyloplasts.

6. Use high power to observe the amyloplasts, and notice that the starch is deposited in layers inside each plastid.

   Make a diagram of a single amyloplast in the following space (Fig. 1.9).

FIGURE 1.9 Individual amyloplast.

Red Pepper (*Capsicum frutescens*) Cells

1. Prepare a wet mount of a small portion of the fleshy part of a red pepper fruit. The region close to the epidermis is best for this.
2. Does the shape of these cells more closely resemble that of the onion or that of the potato? (13)
3. Examine these cells with high power. The small, orange-colored structures are **chromoplasts.** Compare these chromoplasts to the chloroplasts of *Elodea.* (14)

	Chloroplasts	Chromoplasts
Shape		
Size		
Color		

Crystals

Plants produce metabolic waste products just as animals do. However, they lack excretory systems as a way of eliminating these chemicals from their cells or from the entire plant. One way that plants have evolved to escape the harmful effects of the buildup of wastes is to convert them to insoluble substances. These insoluble materials are then stored in the cells without doing any damage. One such harmful waste product is oxalic acid, which is converted to crystals of calcium oxalate. Each of the following crystals is composed of calcium oxalate. Why they differ in shape is not known.

Raphides

1. Cut a very thin cross section of *Zebrina* stem. Place this in a drop of water on a microscope slide.
2. Before adding the cover slip, use a razor blade or dissecting needles to chop up the section. (Raphides have optical properties that are similar to the cytoplasm and cannot be seen inside the cells with our light microscopes.
3. Examine the region where the cells were cut or damaged and notice the long, needlelike crystals that were released from the cells. These elongated crystals are the raphides.
4. Diagram three or four of these raphides in the following space (Fig. 1.10).

FIGURE 1.10 Cluster of raphides from *Zebrina.*

Druses

1. Make a wet mount of a thin cross section of a *Begonia* stem.
2. Look carefully to find the *small* crystals which this plant produces.
3. Diagram two or three *Begonia* druses in the following space (Fig. 1.11).

FIGURE 1.11 Druses from *Begonia.*

Questions

1. How is the total magnification of the compound microscope determined? (15)

2. The dissecting microscope and the compound microscope are designed to serve different purposes. Under what circumstances would you use each microscope? Give examples. (16)

   a. Dissecting microscope:

   b. Compound microscope:

3. List four things that should be done to your compound microscope if it is to be stored properly. (17)

    1.

    2.

    3.

    4.

4. How does the orientation of the image differ when seen through the dissecting and compound microscopes? (18)

    a. Dissecting microscope:

    b. Compound microscope:

5. Why should the coarse adjustment knob *not* be used when the high-power objective is in place? (19)

6. Calcium oxalate is an insoluble precipitate of oxalic acid, which is a strong metabolic toxin. Briefly explain the function or importance of the crystals seen in *Zebrina*. (20)

7. Some plants, for example the jack-in-the-pulpit, have high concentrations of calcium oxalate crystals in their cells. Taking a bite of the raw plant tissue produces a severe burning sensation in the lining of the mouth. Drinking water does not alleviate this pain. Can you suggest a reason why drinking water does not help? (21)

8. How do the druses seen in *Begonia* differ in appearance from the raphides seen in *Zebrina* ? (22)

    a. *Begonia*:

    b. *Zebrina*:

Cell Division and Plant Cell Tissues

Introductory Notes

Plant growth and development involves three main processes: an increase in the number of cells, an increase in the size of the newly formed cells, and differentiation of these cells into mature tissues. The first two events result in an increase in size, while differentiation is responsible for the development of the unique features of the mature plant.

GOALS

After completing this exercise, the student should be able to:

- explain the three different processes involved in plant growth, including a brief discussion of the importance of each step.
- list the stages of the cell cycle in correct order.
- make a labeled diagram of each stage of the cell cycle, and identify each stage from a photograph or diagram.
- explain the difference between mitosis and cytokinesis.
- define a tissue.
- define or describe embryonic tissue, simple tissue, complex tissue, dermal tissue, ground tissue, and vascular tissue.
- identify photographs or diagrams of epidermis, guard cells, stomata, periderm, collenchyma, parenchyma, and sclerenchyma (both sclereids and fibers).
- explain the main function of each of the above.
- answer the questions in the exercise.

THE CELL CYCLE: MITOSIS AND CYTOKINESIS

The formation of new cells in a plant is a process that is generally restricted to specialized regions of embryonic tissue known as **meristematic tissue** (or **meristems**). These are places composed of immature, undifferentiated cells that are capable of repeated cell divisions. The process of cell division consists of two separate but often related events. The first of these is the division of the nucleus and distribution of the genetic material (**mitosis**); the second is the division of the cytoplasm (**cytokinesis**). *The result of mitotic cell division is the formation of two new daughter cells, each containing the same genetic information as the original parent cell.* Through repeated mitotic cell divisions, all of the cells in an organism (except the reproductive cells) have the same number and kind of chromosomes.

This production of new cells in the meristems is a continual process throughout the growth of the plant. For convenience, several steps or stages are usually discussed. In the following exercises, you will examine these stages and learn something about the important events that characterize each stage.

Chromosome Distribution

In this exercise, you will study the movement and distribution of chromosomes as a cell divides by mitosis. The cell you will study will have a diploid (2N) chromosome number of four. (*Diploid* means that there are two of each type of chromosome present in the cell.) Colored pipe cleaners will serve as the chromosomes. The two colors represent chromosomes contributed by the two parents.

Obtain the chromosomes as instructed by your instructor. Also obtain two pieces of paper, labeled Page A and Page B which will serve as the cells involved in the division process.

Follow these instructions and those of your instructor to go through the stages of mitosis.

1. Place four chromosomes (one of each length and color) in the cell on Page A. This will represent the cell in the **interphase** stage of the cell cycle (ignoring the fact that chromosomes are not visible during interphase).
2. The first important event to take place is chromosome **replication**. Each chromosome makes an exact copy of itself. Use the other pipe cleaners to indicate these. Place the other similar pipe cleaners in a cross or X-shaped configuration on the ones present in the cell. The place where two pipe cleaners touch each other will represent the **centromere**. In this state, each pipe cleaner represents a single **chromatid**. This configuration represents **prophase**.
3. Now move the chromosomes through the remaining stages of mitosis.
   a. **Metaphase**. Here the paired chromatids move to the **equatorial plate** of the cell — a region midway between the two poles.

b. **Anaphase.** The chromatids separate from each other at the centromere and move toward the opposite poles of the cell.

c. **Telophase.** The chromosomes stop moving, and two new nuclei will develop. In plant cells, this stage is usually accompanied by the development of the future cell wall that will separate the two new cells. This developing cell wall is referred to as a **cell plate.**

4. This cell plate is the result of **cytokinesis.** Two new daughter cells result. Place the chromosomes onto Page B to show the result.

If you did the procedure correctly, each of the two daughter cells should be identical as far as their chromosome makeup is concerned.

Now repeat the procedure until you can go through it without any prompting from these instructions or from your lab partner. Check each other as a means of testing yourselves.

Diagram the stages of the cell cycle, using the cell outlines in Figure 2.1. A nuclear membrane has been provided for the interphase cell, but you need to show only what happens to the chromosomes. **Keep the original sizes, number, and textures of the interphase chromosomes.**

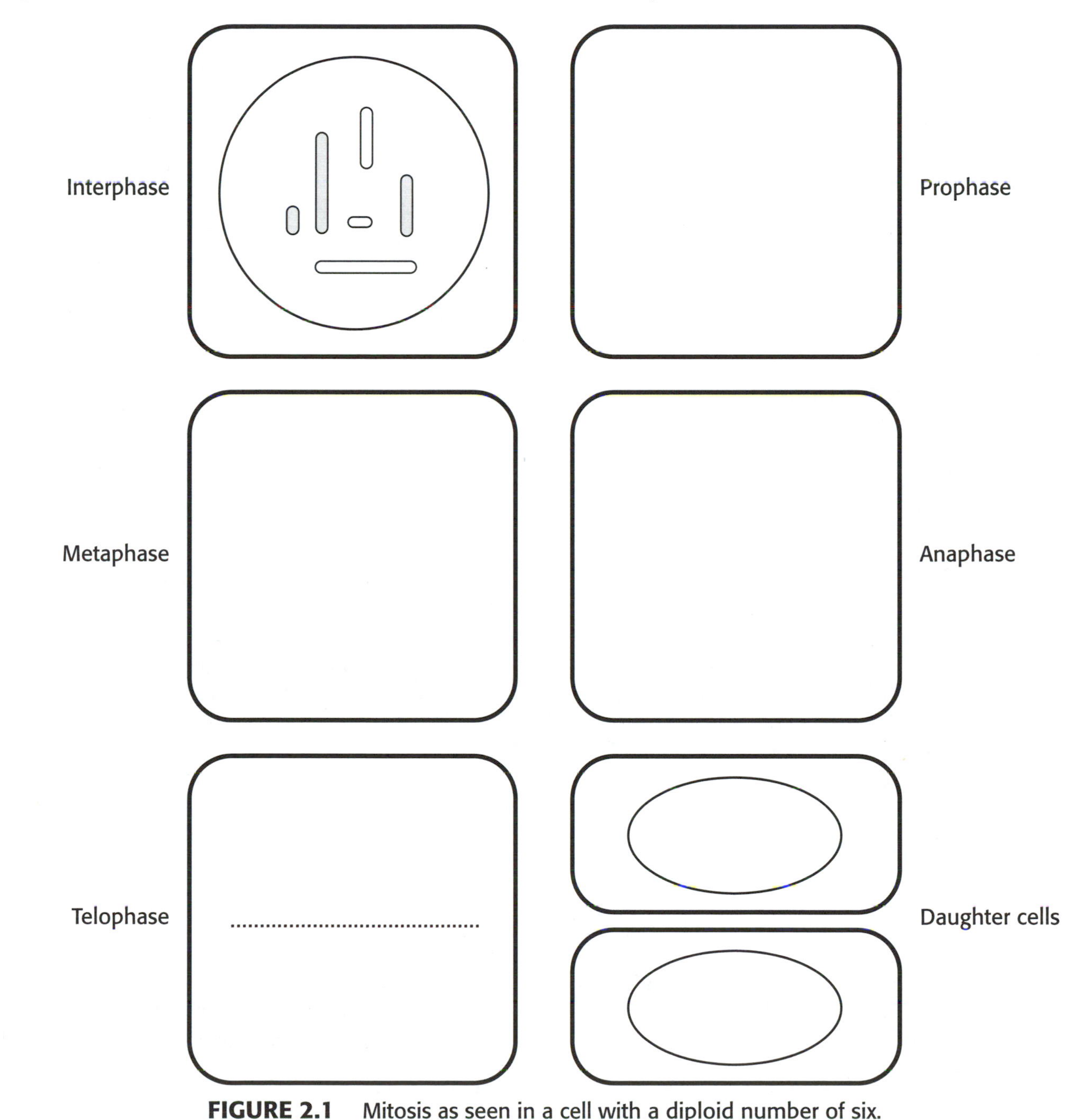

FIGURE 2.1 Mitosis as seen in a cell with a diploid number of six.

Mitosis In Plant Cells

Obtain a prepared slide of a longitudinal section of an onion root tip.

First examine the preparation by using low power. At the extreme tip (the more pointed end) of the root, you should be able to find a group of relatively disorganized cells that form a protective covering of the tip. This is the **root cap**. Do not look here for mitotic stages.

The meristematic region is just above the root cap. Describe these cells. (1)

General shape:

Relative size of the nucleus:

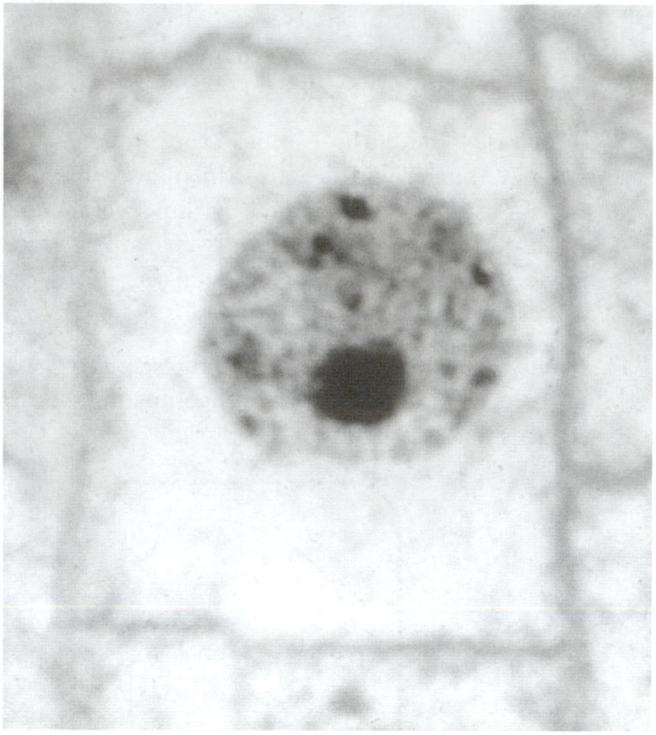

FIGURE 2.2 Onion root tip cell in interphase, X 2330.
Courtesy Carolina Biological Supply Company

The diploid number of chromosomes for *Allium cepa* is 16. Each cell, therefore, has 16 chromosomes. Remember, however, that you are looking at a longitudinal section of the root tip, so parts of chromosomes or even entire chromosomes will have been sliced off during the preparation of the slide.

Now switch to high power and examine the preparation for cells in various stages of the **cell cycle**.

Interphase

This stage is not one of the mitotic phases, but is the longest period of the cell cycle. Most cells will be in this stage. An interphase cell is recognized by its relatively large and intact nucleus. The **nuclear membrane** is intact and its location can be seen with our microscopes. Small, darkly stained **nucleoli** are also clearly visible. The chromosomes are not visible as distinct structures. They are present but are extremely elongated and thin; thus, are visible only as a granular material within the nucleus called **chromatin**.

Label the **nuclear membrane**, **nucleolus**, and **chromatin** in the photograph of the cell in interphase (Fig. 2.2).

Prophase

The chromosomes first become visible during the early stages of prophase. This change in visibility is due to a shortening and thickening of the chromatin. Also during early prophase, the nuclear membrane and the nucleoli begin to deteriorate.

By late prophase, the nucleoli will have nearly disappeared, and the nuclear membrane cannot be seen. The chromosomes will be much thicker and can be seen as being composed of two parallel threads called **chromatids**. The two chromatids are held together by a specialized region known as the **centromere**, which cannot be seen on our slides.

This paired condition of the chromosomes is the result of **replication** (duplication) of the genetic material, which happens during interphase. The duplication is not evident until the chromosomes become more condensed (shorter and thicker). This process of genetic replication is of vital importance for the transfer of genetic information from a parent cell to the two resulting daughter cells.

Label the **chromatids** and **nucleolus** in the photograph of a cell in early prophase (Fig. 2.3).

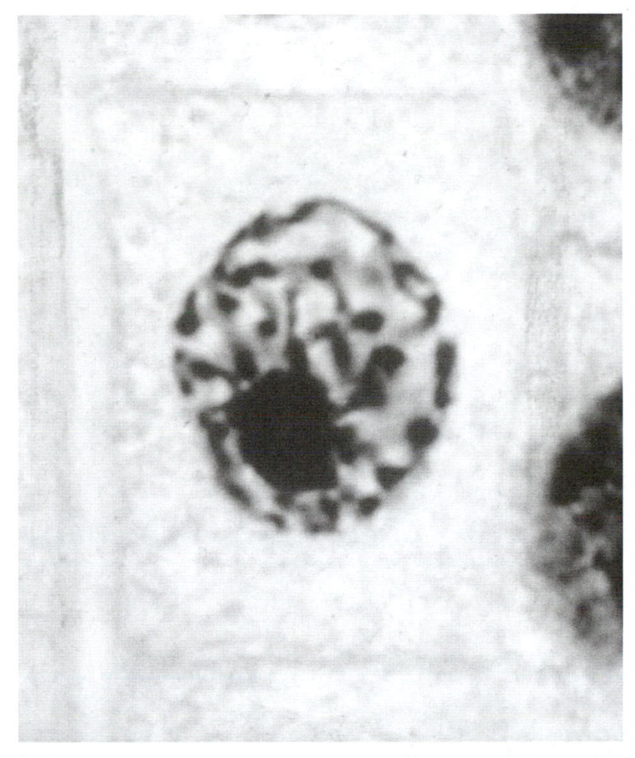

FIGURE 2.3 Onion root tip cell in early prophase, X 2330. Courtesy Carolina Biological Supply Company.

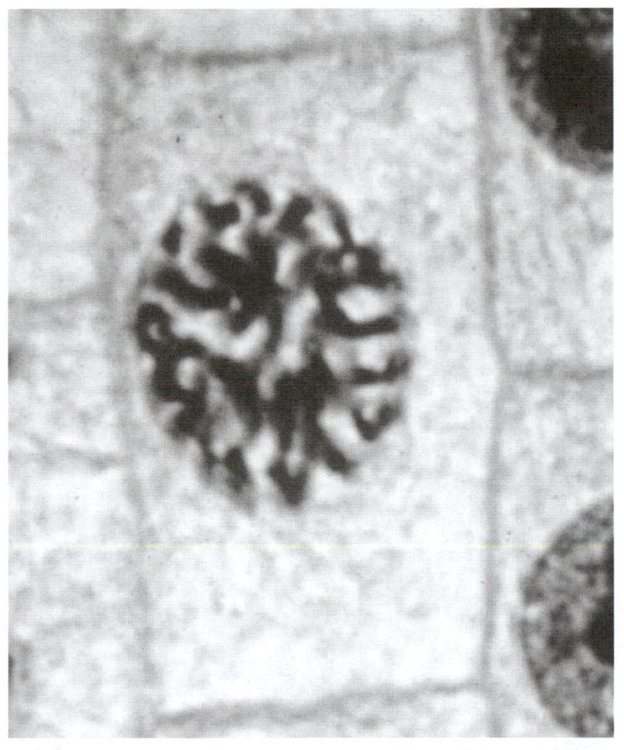

FIGURE 2.4 Onion root tip cell in later prophase, X 2330. Courtesy Carolina Biological Supply Company

Metaphase

This stage is easily identified by the arrangement of **chromatids** along the equatorial plate of the cell. By this time, the nucleoli and nuclear membrane have completely disappeared. Very thin, threadlike structures called **spindle fibers** will have formed. Some spindle fibers extend from one **pole** of the cell to the other, while others extend only from a pole to a centromere on a chromosome. Spindle fibers are very difficult to find; be patient and use different light intensities.

Label the **spindle fibers, polar region,** and **chromatids** in the cell in metaphase of mitosis (Fig. 2.5).

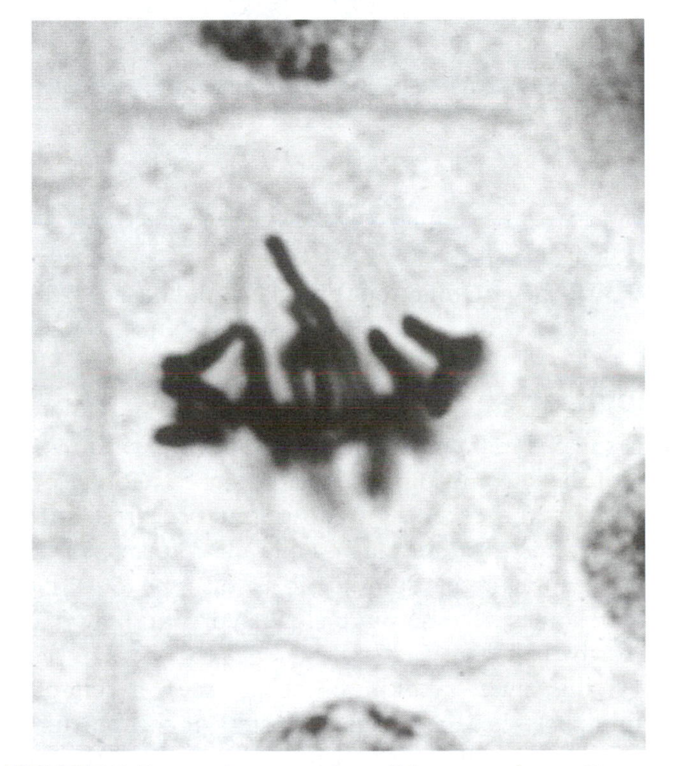

FIGURE 2.5 Onion root tip cell in metaphase, X 2970. Courtesy Carolina Biological Supply Company

Anaphase

This stage is characterized by the separation of the chromatids (beginning first at the centromeres). After separation, each chromatid is known as a **chromosome.** The two separate chromosomes then move toward opposite poles of the cell. Somehow, the spindle fibers are involved in this process. They may actually pull the chromosomes toward the poles. In an anaphase cell, two clusters of chromosomes can be seen.

Label a **chromosome** and **polar region** in the photograph of a cell in anaphase of mitosis (Fig. 2.6).

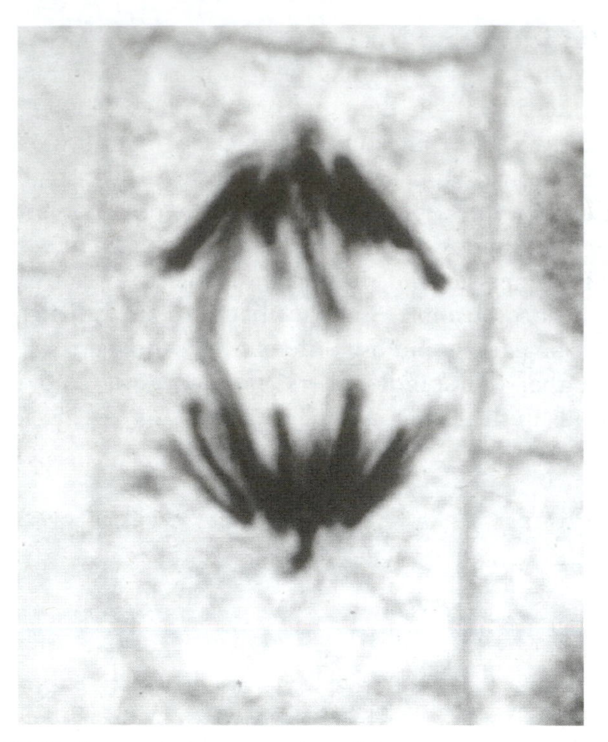

FIGURE 2.6 Onion root tip cell in anaphase, X 2560.
Courtesy Carolina Biological Supply Company.

Telophase

This stage is marked by the end of the poleward migration of the chromosomes. Other events of telophase are essentially the reverse of those that occurred during prophase, i.e., the nuclear membrane reforms, the nucleoli reappear, the spindle fibers disappear, and the chromosomes become more elongated. Two new nuclei are formed.

At this time, **cytokinesis** occurs with a **cell plate** forming between the two new clusters of **chromosomes.** The cell plate is the early form of the cell wall that will separate the two resulting cells.

Does the formation of the cell plate begin in the middle of the cell or more toward the periphery? (2)

Label the **chromosomes** and **cell plate** in the photograph of a cell in telophase of mitosis (Fig. 2.7).

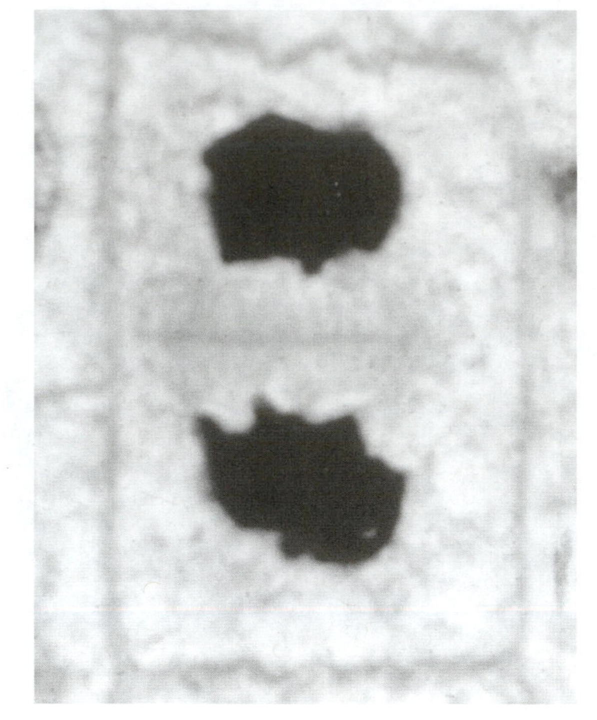

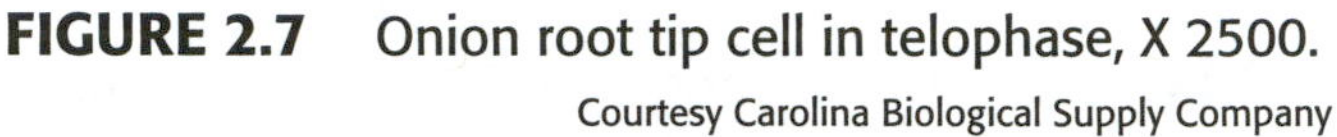

FIGURE 2.7 Onion root tip cell in telophase, X 2500.
Courtesy Carolina Biological Supply Company

Daughter cells

The two cells that result from this process are known as daughter cells. How can you recognize such newly formed cells? (3)

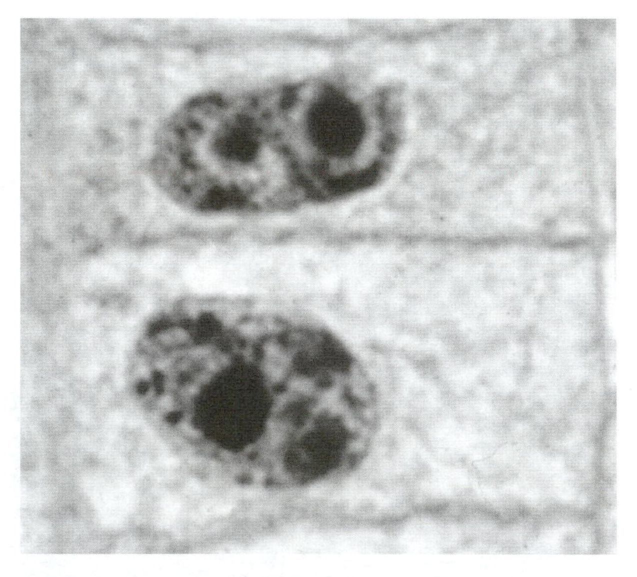

FIGURE 2.8 Daughter cells in an onion root tip after completing the cell cycle, X3600.
Courtesy Carolina Biological Supply Company.

PLANT TISSUES

The new cells formed in the meristems of a plant are all similar in appearance and even function. Some of these cells must, by necessity, remain meristematic. Most, however, mature and become associated with other cells to form tissues. A tissue may be composed of one or more types of differentiated cells and usually performs a specific function.

The tissues of higher land plants may be classified in the following manner:

I. Embryonic tissue (meristems)

II. Permanent tissues

- A. Dermal or surface tissues
  1. Epidermis
  2. Periderm (cork)
- B. Simple fundamental tissue (ground tissue)
  1. Parenchyma
  2. Collenchyma
  3. Sclerenchyma
  4. Endodermis
- C. Conducting tissue (vascular tissue)
  1. Xylem
  2. Phloem

A **simple tissue** is one that is composed of only a single cell type. These cells are mature and normally have no meristematic activity. **Complex tissues** contain more than one cell type. The xylem and phloem are complex tissues and will be studied in more detail in other lab exercises.

Meristematic Tissue

There are two principle types of meristems found in higher plants: apical meristems and lateral meristems. Apical meristems are located at the tips of stems and roots. The activity of these meristems results in an increase in the length of the plant. This is known as **primary growth**, and the resulting tissues are called **primary tissues.** Lateral meristems are found near the periphery of stems and roots and are responsible for an increase in diameter. This in known as **secondary growth** (and **secondary tissues**). These will be covered in later exercises.

Surface Tissues

Epidermis

This is the most important surface tissue of young plants, leaves, and herbaceous plants that lack lateral meristems. Frequently, it is only a single cell layer thick.

1. Obtain a leaf from a *Zebrina* plant.
2. Tear the leaf crossways, i.e., at a 90° angle to the veins. Look for a small region of the **lower** (purple) epidermis to show at the torn margin.
3. Use distilled water and make a wet mount of this purple tissue.
4. First examine the preparation with low power. You should see two types of cells: the true **epidermal cells** and **guard cells.**
5. Switch to high power and observe some guard cells. Look carefully and you will see two guard cells; they always occur in pairs, side by side. Describe how the guard cells differ in *shape* from the surrounding epidermal cells. (4)

   Shape of single guard cell:

   Other epidermal cells:

   How does the interior of the guard cells differ from that of the surrounding epidermal cells? (5)

   Guard cells:

   Other epidermal cells:

   The small space between the two guard cells is known as a **stoma** (pl. = **stomata**). Notice the lack of air spaces between the other epidermal cells, and also the lack of chloroplasts in these cells.

6. What are the *functions* of the two types of cells in the *Zebrina* leaf epidermis?

   The epidermal cells protect against (6)

   The guard cells (7)

7. The lower epidermis of *Zebrina* is a complex tissue. Why? (8)

8. Figure 2.9 is a photograph of the lower epidermis of a *Zebrina* leaf. Label a **guard cell**, the **stoma**, an **epidermal cell**, and a **chloroplast**.

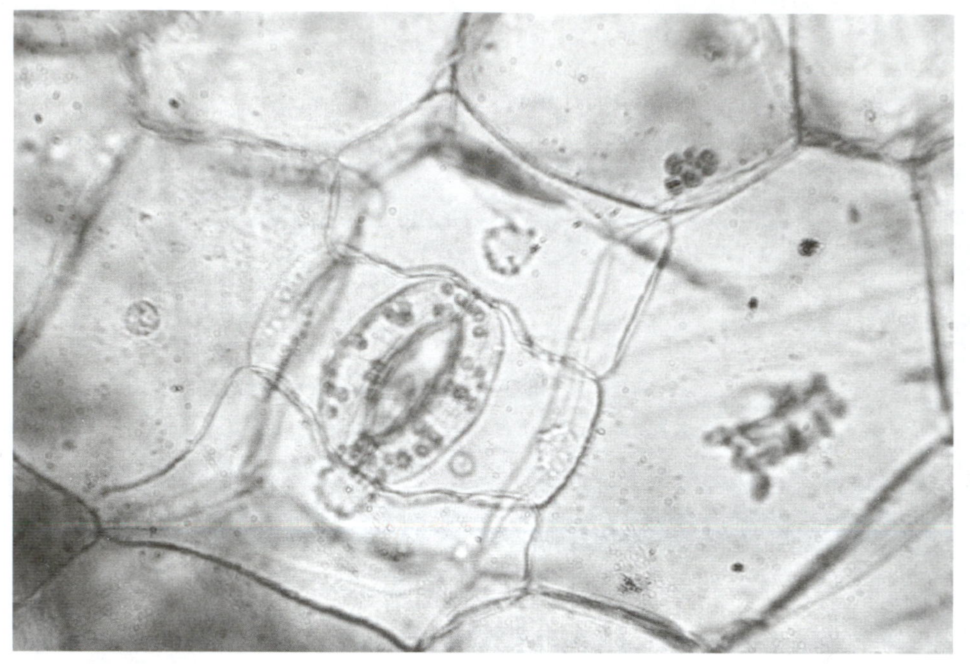

FIGURE 2.9 Lower epidermis of *Zebrina* with guard cells, X 320.

Periderm or cork

In Exercise 1, you examined cork tissue. Cork comprises the most abundant part of the periderm tissue in woody plants. The periderm replaces the epidermis on those plants that have active lateral meristems and is usually several to many cells thick. Cork cells fit together very snugly and are dead at functional maturity, giving this tissue no nutritive value; both are important features when considering the function of this tissue. If you feel it necessary, repeat the observation. Instructions are in Exercise 1, on page 6. Corks are available.

Simple Fundamental Tissues

Collenchyma

Collenchyma is a simple tissue composed of elongated cells with tapered ends. In cross section, you should see small cells with **irregularly thickened cell walls** (thicker in the regions where three or four cells are in contact).

1. Make a wet mount of a very thin *cross section* of a celery (*Apium graveolens*) petiole. Take your tissue from the outer edge of the stalk as shown.

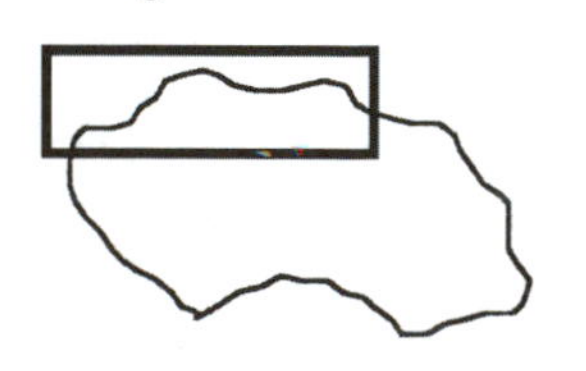

2. With low power, locate the small clusters of light gray cells that lie immediately adjacent to the epidermis on each of the prominent ridges along the outer portion of the petiole. These are strands of collenchyma tissue, comprising the "celery strings."
3. If necessary, add a drop of 0.1% neutral red dye next to the edge of the cover slip, and allow the dye to flow under the cover slip. Use a piece of paper towel at the opposite side of the cover slip to draw the dye under faster. Neutral red will stain the the cell walls.
4. After a few minutes, observe your preparation again. This time, use high power to see the characteristic irregular thickenings of the cell walls.
5. What is the *function* of collenchyma tissue? (9)

Parenchyma

Parenchyma is the most common kind of simple tissue. It is found in roots, stems, leaves, flowers, and fleshy fruits. This type of tissue makes up the bulk of the herbaceous plant body.

1. The same section of celery that was used above for observation of collenchyma can be used here. Examine the larger cells in the center of the petiole.

2. Notice the **thin cell walls**, generally isodiametric shape of the cells, and the **intercellular spaces** (air spaces where three or four cells are in contact). These are the general characteristics of this type of tissue.

Sclerenchyma

This tissue differs from the others considered thus far in that sclerenchyma cells have thick secondary cell walls and are dead at functional maturity. The cell walls of sclerenchyma cells are thick, hard, and contain lignin. In contrast to collenchyma, sclerenchyma tissue has cells with **evenly thickened cell** walls. There are two general types of sclerenchyma cells: elongated **fibers** and **sclereids** which are not elongated, but may come in a variety of shapes.

Sclereids

1. Prepare a wet mount of a *small amount* of the fleshy part of a pear (*Pyrus communis*) fruit. The small, gritty particles that you feel when you eat a pear are clusters of **sclereids** (often called **stone cells**).
2. Examine this preparation with low power, and notice that the clusters vary in size, from only a few cells to 50 or more.
3. It is difficult to see the individual cells in these clusters, so take a cork, or use the eraser of your pencil, and apply pressure to your cover slip. This will separate the cells.
4. Now locate two sclereids that are still in contact with each other. Examine these with high power.
5. Describe the general shape of *a single* pear sclereid. (10)
6. How does the thickness of the cell wall compare to that of other cells studied thus far? (11)
7. Notice the cell cavity or **lumen** and the numerous **pit canals** that radiate through the cell wall from the lumen. The lumen is where the main cell body and nucleus once existed, and the pit canals mark places where strands of cytoplasm connected two adjacent cells. In living cells, these strands of cytoplasm are known as **plasmodesmata**. These are dead cells, remember, and all that can be seen are the tubes where the plasmodesmata once were.
8. Draw two adjacent pear sclereids. Show and label the **pit canals** connecting the two cells, and the **lumen** of each cell (Fig. 2.10).

FIGURE 2.10 Individual pear sclereids.

Fibers. Fibers are elongated sclerenchyma cells with thick cell walls and small lumens. They are often found in association with vascular bundles, forming a bundle cap of cells just outside the phloem. In some cases, they may completely encircle the vascular bundle.

1. Obtain a prepared slide of either alfalfa (*Medicago* sp.) or sunflower (*Helianthus* sp.) stem. Examine the preparation with low power. The clusters of cells that form a ring around the periphery of the stem are vascular bundles.
2. Toward the outside of each vascular bundle is a cluster of sclerenchyma fibers. Examine these with high power and note the thick cell walls and small central lumens.

   Label the **sclerenchyma fibers** in Figure. 2.11.

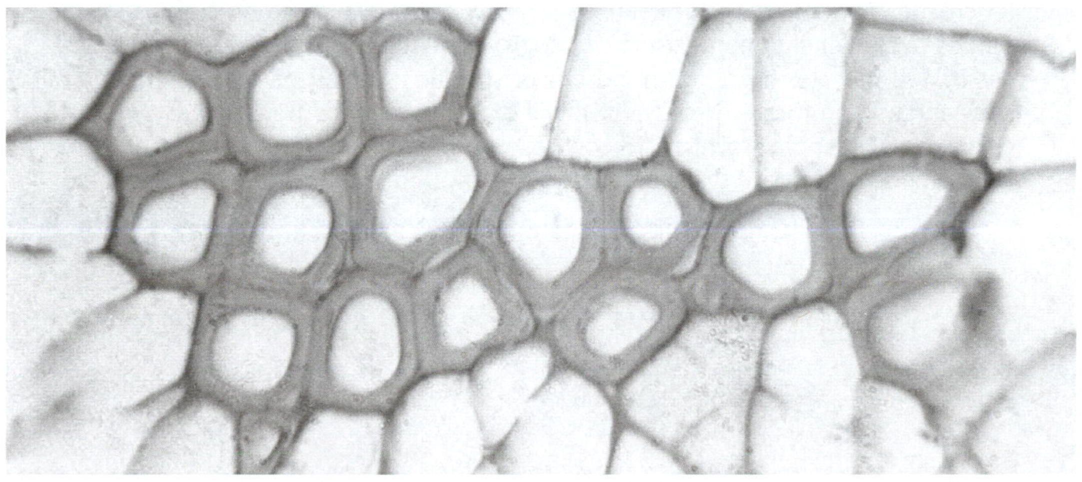

FIGURE 2.11 Cross section of sclerenchyma fibers found in a *Coleus* stem, X 550.

3. What four tissues, studied today, can be seen on the slide of *Medicago* or *Helianthus*? (12)

4. Label these four tissues in Figure 2.12, below.

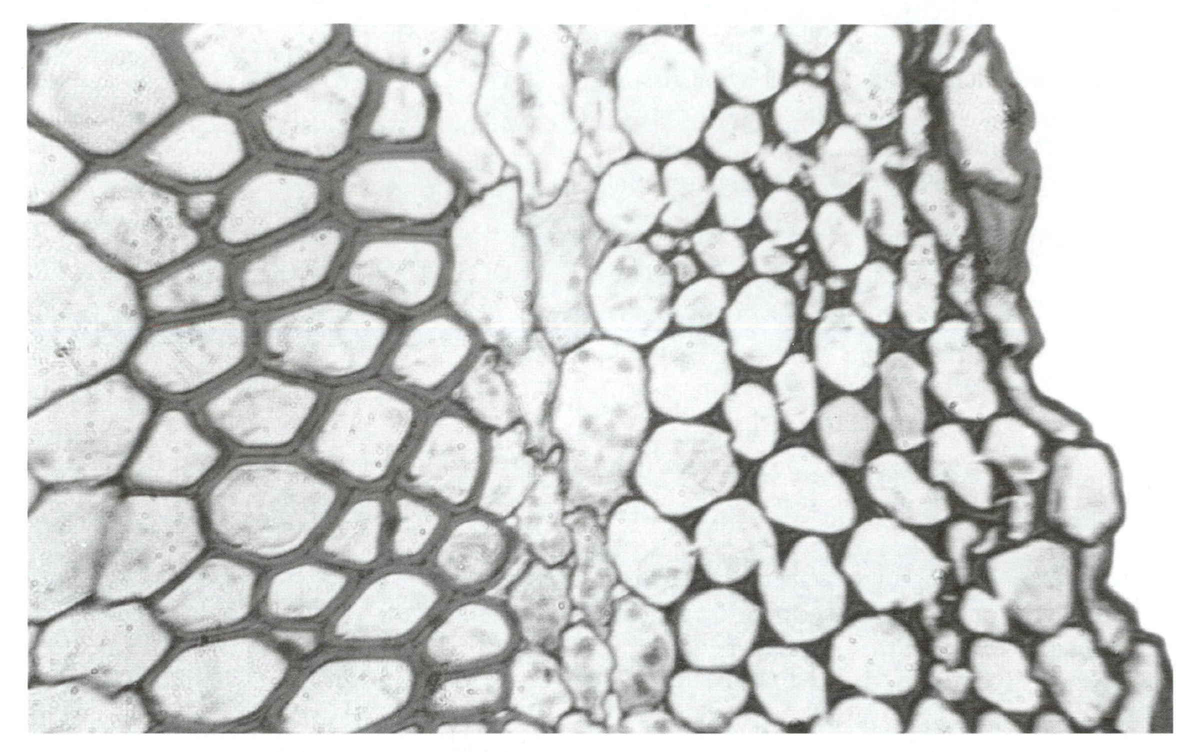

FIGURE 2.12 Simple fundamental plant tissues, X 385.

IMPORTANT NOTE: Parenchyma, collenchyma, and sclerenchyma are basic cell types that often form simple tissues. Frequent references will be made to these three tissues. Know their characteristics and main functions. Parenchyma, collenchyma, and sclerenchyma are basic tissue types that have more specific names. A similar situation exists in zoology where there is a type of tissue known as striated muscle. There are hundreds of specific muscles, each having its own special name. In future exercises, we will examine plant tissues such as the pith, cortex, and mesophyll, which are composed primarily, or entirely, of parenchyma cells. Henceforth, these three tissues will be referred to as tissue **TYPES**.

Example: 1. **Q.** What tissue is found in the center of a clover stem?
A. The pith

2. **Q.** What **type** of tissue is found in the center of a clover stem?
A. Parenchyma

Questions

1. What is the important result of mitosis? (13)

2. Mitosis is often considered synonymous with cell division. Why is this erroneous? (14)

3. Describe a cell in each of the following stages of the cell cycle. Give the main feature(s) you would use to identify the various stages of mitosis. (15)

   a. Interphase:

   b. Prophase:

   c. Metaphase:

   d. Anaphase:

   e. Telophase:

4. What is a tissue? (16)

5. What are some identifying features of meristematic tissue? (17)

   a. Relative size of the cells:

   b. Relative size of the nucleus:

   c. Size of the vacuoles:

6. What is the *function* of periderm? (18)

7. What are two *functions* of parenchyma (other than giving structure to the plant)? (19)

8. Collenchyma and sclerenchyma fibers both have a similar main *function*. What is it? (20)

9. What is a possible *function* of the sclereids found in the pear fruit? (21)

10. In Exercise 1, you examined potato tuber cells. What *type* of tissue were you examining? (22)

11. Find and **label *each stage* of the cell cycle** seen in Figure 2.13

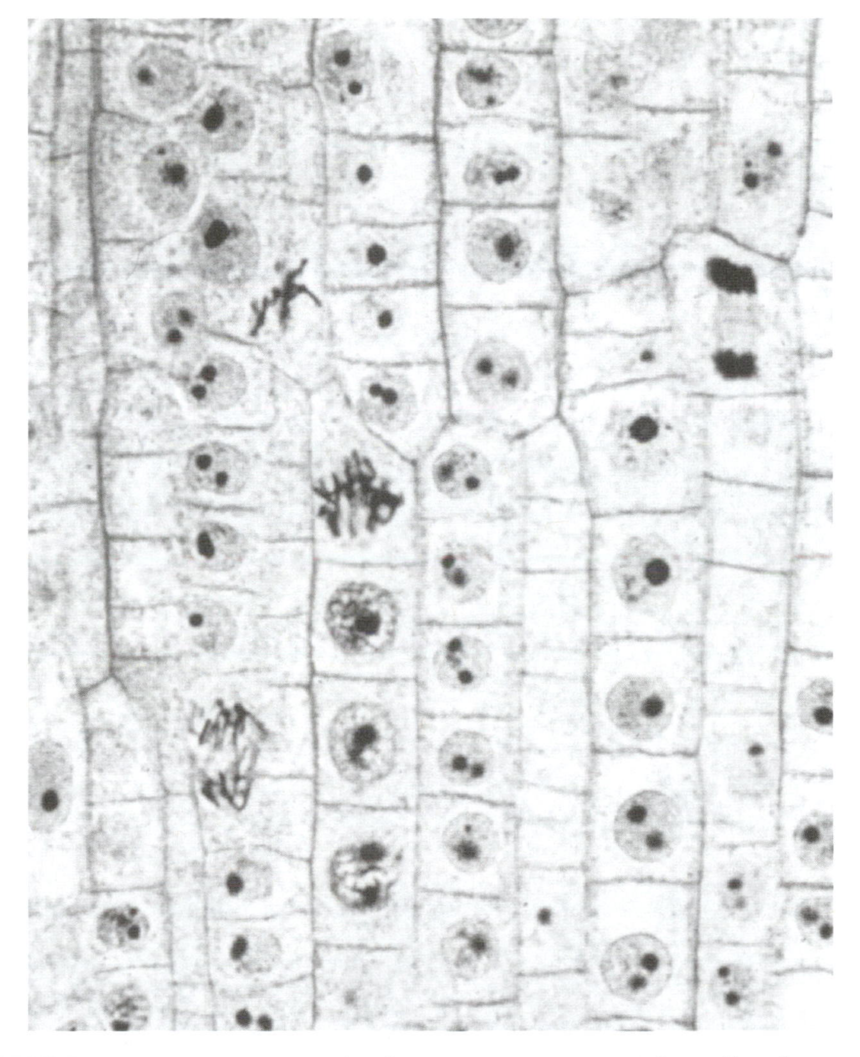

FIGURE 2.13 Onion root tip cells in various stages of the cell cycle, X 570.

Courtesy Carolina Biological Supply Company

Stems: Herbaceous Plants

Introductory Notes

The stem is that portion of the plant that supports the leaves, flowers, and fruits. There is a great deal of variety in stem structure and function. This exercise deals with some attributes of aerial stems of herbaceous plants. Herbaceous plants are those that usually have moderately soft, succulent, green stems and lack deposits of woody tissues. Typically, these plants die or at least die back at the end of the growing season.

This exercise and the one following deal with typical aerial stems. You should be aware that there are many stem modifications. Your text describes some of these (bulbs, corms, rhizomes, etc.).

GOALS

After completing this exercise, the student should be able to:

- list four functions of plant stems.
- explain what is meant by primary growth and primary tissues.
- list the three primary meristems of a stem tip.
- be able to identify, from either a photograph or diagram, the three primary meristems of a stem tip and explain which mature tissues will develop from each.
- explain how the vascular bundles are arranged in a typical dicot and how this differs from that found in a typical monocot stem.
- identify, from a photograph, the main tissues of an herbaceous dicot stem and of a monocot stem.
- describe the main components of the phloem, i.e., sieve tube members and companion cells.
- explain the difference between primary and secondary growth.
- explain how a stem grows in both length and diameter.
- define, and/or recognize from a photograph, the following: axil, node, apical meristem, protoderm, ground meristem, procambium, leaf trace, leaf gap, leaf primordium, bud primordium, pith, cortex, epidermis, bundle cap, primary growth, primary phloem, primary xylem, sieve tube member, sieve plate, companion cell, dicot stem, monocot stem, bundle sheath, secondary growth, secondary phloem, secondary xylem, fascicular cambium, interfascicular cambium, and pith ray.
- answer the questions in the exercise.

EXTERNAL MORPHOLOGY

Two very common plants will be used to serve our purpose here: *Coleus* and the florist's geranium (*Pelargonium*). One shows **alternate** leaf arrangement, and the other has **opposite** leaf arrangement.

The *region* of the stem where the leaves arise is known as the **node**.

How many leaves are found at each node on the *Coleus* plant? (1)

What term is used to describe this leaf arrangement? (2)

How many leaves are found at each node in *Pelargonium* and what is the term used to describe this arrangement? (3)

The angle between the stem and the upper leaf surface is called the **axil**.

What is found in the leaf axil near the top of the plant? (4)

What is found in the axil region of more mature leaves nearer the base of the plant? (5)

THE STEM TIP AND PRIMARY GROWTH

Primary growth refers to an increase in the length of a stem or root. Both stems and roots increase in length only at their tips. At the very tip of the stem is a region of active cell division, the **apical meristem**. Below this is a region of cell elongation and finally a zone of cell maturation.

Obtain a prepared slide of a longitudinal section of a *Coleus* stem tip. Here you will be able to observe the three primary tissues and early stages of stem development.

1. Locate the **leaf primordia** and **bud primordia**, the rudimentary structures that will develop into leaves or buds (sing. = **primordium**). Your observation of the entire plant (Part B) should help you determine which is which. Immediately beneath the apical meristem, three primary meristems develop. These are (1) the **protoderm**, which will develop into the **epidermis**, (2) the **procambium**, which gives rise to the **primary vascular tissue**, and (3) the **ground meristem**, which will mature into the **pith** and **cortex** in dicots. The cortex is that tissue between the epidermis and the vascular bundle. The pith is the parenchyma tissue in the center of the stem.

2. Pay particular attention to the procambium. Notice that traces of this tissue extend out into the leaf primordia. A vascular bundle that connects the vascular system of a leaf to that in the stem is referred to as a **leaf trace**. Where a leaf trace arises, it leaves a space above it in the stem's vascular cylinder that becomes filled with parenchyma tissue. This region of parenchyma cells is referred to as a **leaf gap**. Leaf traces and leaf gaps are found in the stem.

   Figure 3.1 is a photograph of a longitudinal section of a *Coleus* stem tip. Label the following parts: **apical meristem, leaf primordium, bud primordium, protoderm, procambium, ground meristem, leaf trace, leaf gap, pith, cortex, epidermis,** and maturing **vascular tissue.** Label the mature tissues at the bottom of the photograph.

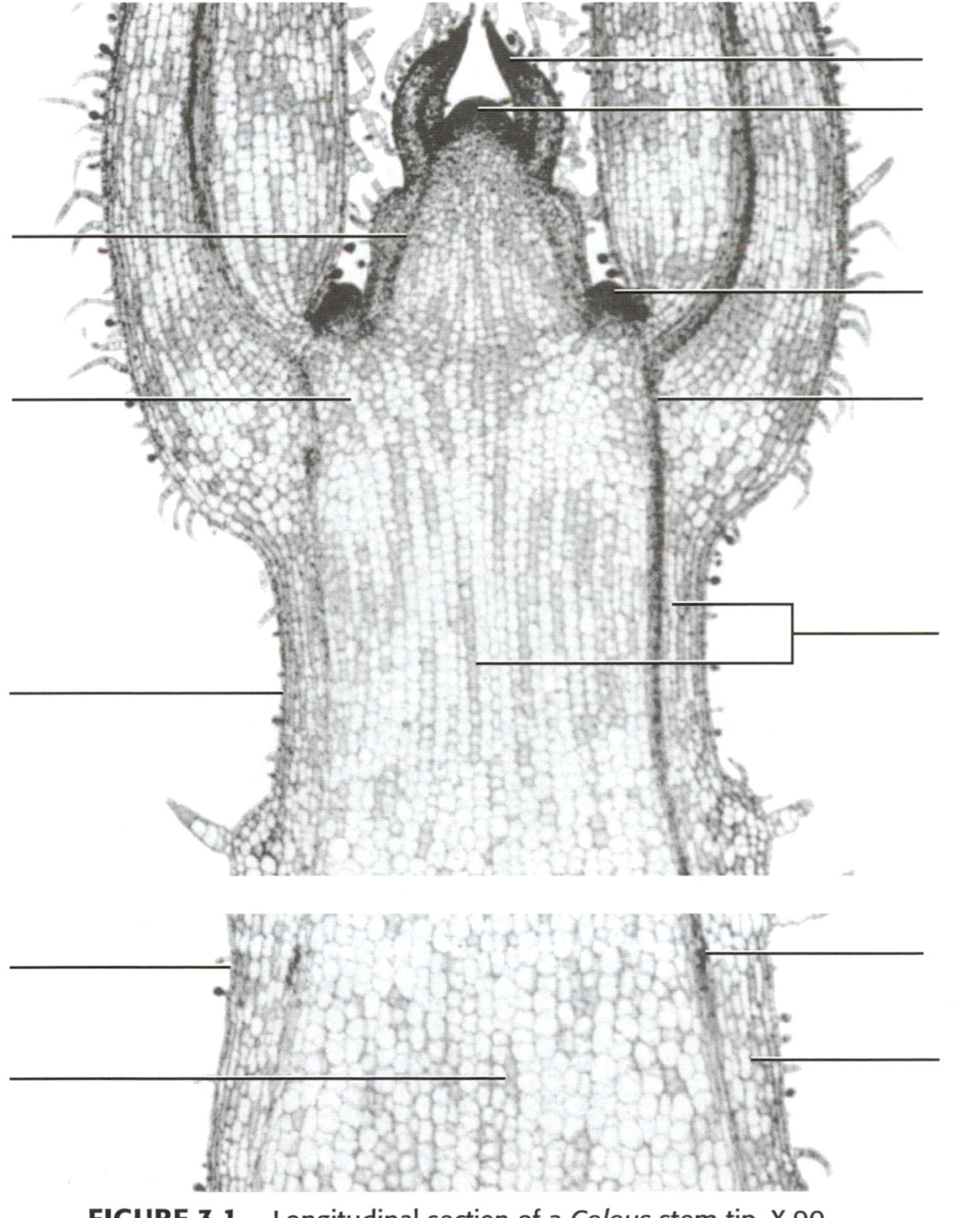

FIGURE 3.1 Longitudinal section of a *Coleus* stem tip, X 90.

PLANTS HAVING ONLY PRIMARY TISSUE

Corn (*Zea mays*), a Monocot

Examine a prepared slide of a cross section of a *Zea mays* stem with the dissecting microscope. When studying plant anatomy, it is all too easy to concentrate on identifying the various cells and tissues and forget that these are part of an integrated system. You should acquire the habit of using both microscopes; one for tissue examination and one to get an idea of the overall organization of these tissues.

The vascular tissues (xylem and phloem) can be seen as small clusters of cells. Use Figure 3.2 to show the *distribution* of the vascular bundles. Do not show any cellular detail. Simply use small ovals to show where most of the vascular bundles are located.

This arrangement of vascular bundles is typical of monocots. Because of this arrangement, no distinction can be made between cortex and pith. The tissue comprising most of the stem is simply referred to as **ground tissue.** This is a mature tissue and should not be confused with the ground meristem.

Now examine a singular vascular bundle with high power. Orient your slide in such a manner that the vascular bundle resembles a face. The red stained cells that outline the vascular bundle are **sclerenchyma fibers,** which form a **bundle sheath.** On some of the slides, this tissue may not be completely developed. The green "forehead" is the **primary phloem,** in which you should be able to distinguish two types of cells. The larger ones are **sieve tube members,** and the smaller ones are **companion cells.** The "eyes" and "nose" are large **vessel elements** of the **primary xylem.** The "mouth" is an **air space** formed by the stretching and tearing of some of the first-formed xylem cells. It probably has no function.

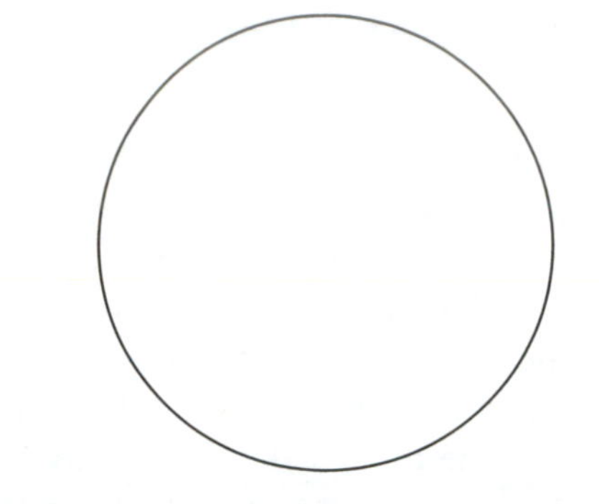

FIGURE 3.2 Distribution of vascular bundles in a corn stem.

Examine the epidermal region of the stem, and notice that the vascular bundles are closer together here. There are relatively more schlerenchyma fibers in this region and the parenchyma cells have thicker cell walls. This is the rind and contributes much to the mechanical support of the stem.

Figure 3.3 is a single vascular bundle of corn. Label the **sclerenchyma fibers,** surrounding **ground tissue, sieve tube member, companion cell, large vessel element, primary xylem,** and **air space.**

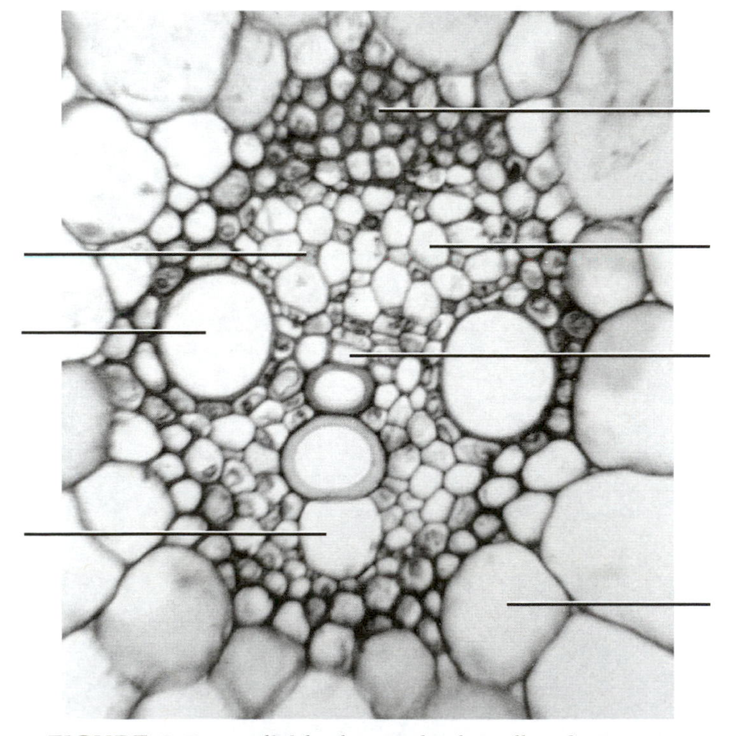

FIGURE 3.3 Individual vascular bundle of corn, X 345.

Buttercup (*Ranunculus*), a Dicot

Examine a prepared slide of a cross section of a *Ranunculus* stem with the dissecting microscope. Use Figure 3.4 to show the *distribution* of the vascular bundles. As you did for the corn stem, do not show any cellular detail. Simply use ovals to show where the vascular bundles are located. Notice that *Ranunculus* has a hollow stem. This central space will be referred to as a **pith cavity**.

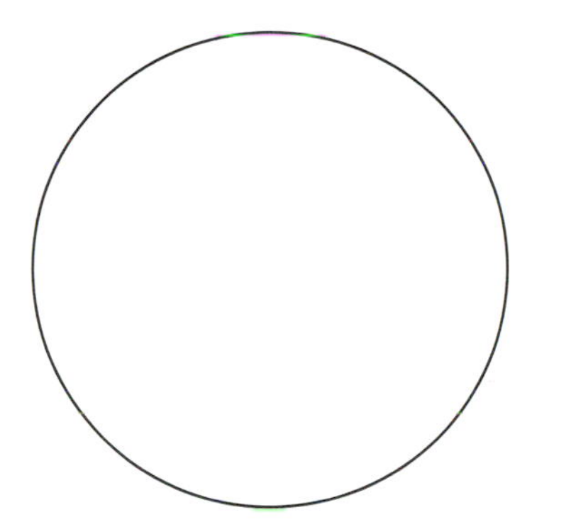

FIGURE 3.4 Distribution of vascular bundles in *Ranunculus.*

What is the most abundant type of tissue in the *Ranunculus* stem? If necessary, use the compound microscope to determine this. (6)

Now use the compound microscope to examine a vascular bundle in more detail. **Sclerenchyma fibers** may be present in some bundles. If so, they will appear as red-stained cells with thick cell walls at the outer portion of the bundle. The smaller cells in the center of each bundle comprise the **primary phloem**. In the *Ranunculus* stem, it is difficult to recognize the sieve tube members and the companion cells that make up the phloem tissue.

The **primary xylem** is immediately adjacent to the primary phloem toward the center of the stem. The cells here are of two types: (1) the larger and more regularly arranged **vessel elements** and (2) the smaller **xylem parenchyma.**

Notice the chloroplasts in the cells of the cortex. Parenchyma tissue such as this is often referred to as **chlorenchyma.**

Figure 3.5 is a portion of a *Ranunculus* stem. Label the **epidermis, cortex, sclerenchyma fibers, primary phloem, primary xylem, vessel element, chlorenchyma tissue, guard cells with stoma, pith,** and **pith cavity.**

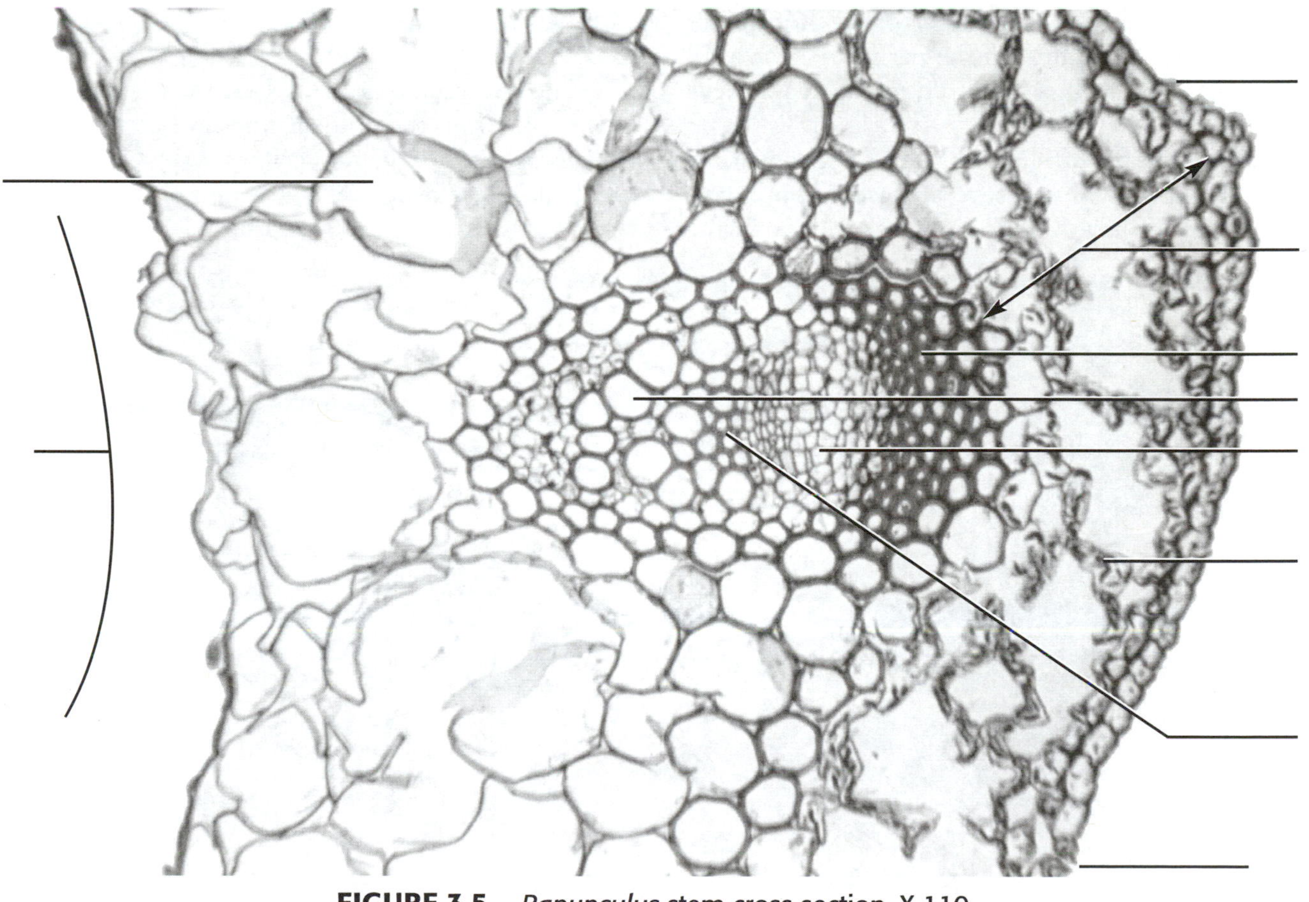

FIGURE 3.5 *Ranunculus* stem cross section, X 110.

Zebrina, a Monocot

Make a wet mount of a *longitudinal section* of about 1 cm of a *Zebrina* stem. After you have added the cover slip, apply some pressure to flatten the preparation slightly. Examine the stem for some xylem vessels. These will appear as elongated tubes with spiral or ringlike (annular) thickenings. **Note:** *a* ***vessel*** *is composed of a series of vessel elements, end-to-end.*

Diagram these thickenings in the outline of a portion of a *Zebrina* vessel (Fig. 3.6).

FIGURE 3.6 A portion of a vessel from a *Zebrina* stem.

HERBACEOUS STEMS WITH SOME SECONDARY GROWTH

Squash or Pumpkin (*Cucurbita*), a Dicot

This plant is used here to examine the **sieve tube members** and **companion cells** of the **primary phloem**. Because the sieve tube members are relatively short, there is a high probability of finding the perforated end wall of these cells. This cell wall is referred to as a **sieve plate.**

Examine a prepared slide of a cross section of a squash stem. You will see several vascular bundles. Each of these will have very large xylem vessels in the center and will have primary phloem on both the outer and inner part of each bundle (a change from the usual arrangement of the phloem toward the outside with the xylem toward the inside).

Find the very large **vessel elements** in the primary xylem. Look in the phloem for **sieve plates, sieve tube members,** and **companion cells.**

Label these structures in the photograph of a squash stem cross section (Fig. 3.7).

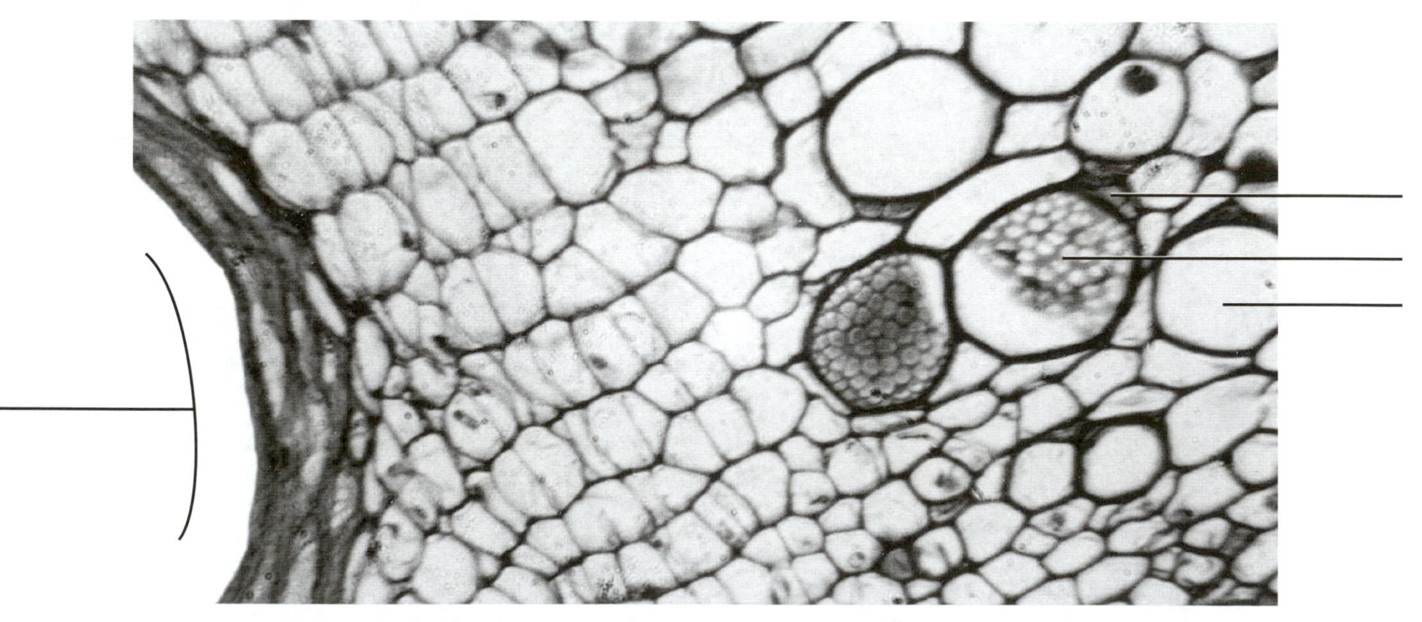

FIGURE 3.7 Cross section of a squash stem, X 360.

The Sunflower (*Helianthus*)

In the previous examples, all of the procambium differentiated into either primary xylem or primary phloem. Such plants generally do not grow very large. In many plants, however, some of the procambium remains meristematic. This meristematic tissue is called the **vascular cambium** and is located between the primary xylem and primary phloem. The cell divisions that occur in the cambium result in an *increase in diameter* of the stem or root. Tissues produced by the vascular cambium are called **secondary tissues**. The resulting increase in size is called **secondary growth**. Those cells formed toward the interior of the stem comprise the **secondary xylem**, while those produced toward the outside make up the **secondary phloem**. Notice that the primary xylem and primary phloem eventually become located at the extreme inside and outside of each vascular bundle. The vascular cambium is best developed in woody plants, producing secondary vascular tissue for as long as the plant lives. Most herbaceous plants have a cambium that functions for only a short time, and secondary tissues are not very prominent. Monocots lack a vascular cambium.

Obtain a prepared slide of a cross section of a *Helianthus* stem. Examine it first with the dissecting microscope. Notice how the distribution of the vascular bundles is similar to that seen in the *Ranunculus* stem. This cylindrical arrangement of vascular tissue is a general characteristic of dicots.

Use a compound microscope to find the following tissues. Start from the outside of the stem.

1. The **epidermis** is the single layer of cells at the outside region of the stem. Stomata are present here but are not abundant and probably closed.
2. The large cluster of red-stained cells at the outer part of each vascular bundle are **phloem fibers**, often referred to as the bundle cap.
3. The tissue between the epidermis and the vascular bundle, not including the bundle cap) is the **cortex**. What two types of tissues can be seen here? (7)
4. The **primary phloem** can be seen in the vascular bundle, immediately adjacent to the bundle cap.
5. The **secondary phloem** is not well developed on these slides.
6. The **vascular cambium** will appear as a line of thin-walled, green cells going around the stem.
   a. What kind of tissue is this? (8)

   b. What is its function? (9)

Notice that the cambium forms a complete circle around the stem. The cambium tissue found *within* a vascular bundle is the **fascicular cambium**. The cambium that develops *between* two adjacent vascular bundles, connecting them, is called the **interfascicular cambium.**

7. The **secondary xylem** is also probably not well developed. Any secondary xylem tissue will be found immediately to the interior of the fasicular cambium.
8. The **primary xylem** is the tissue with large, red-stained cells that are more or less in straight lines. Both **vessel elements** and parenchyma cells are present.
9. The tissue found in the central portion of the stem is the **pith**. What *type* of tissue is this? (10)
10. The extensions of the pith, *separating* the individual vascular bundles, are known as **pith rays**. Similar, but smaller, rays can be found in woody plant stems.
11. Label the **epidermis, cortex, phloem fibers, primary phloem, interfascicular cambium, fascicular cambium, secondary xylem, primary xylem, pith, pith ray**, and a **vessel element** as found in a sunflower stem (Fig. 3.8).

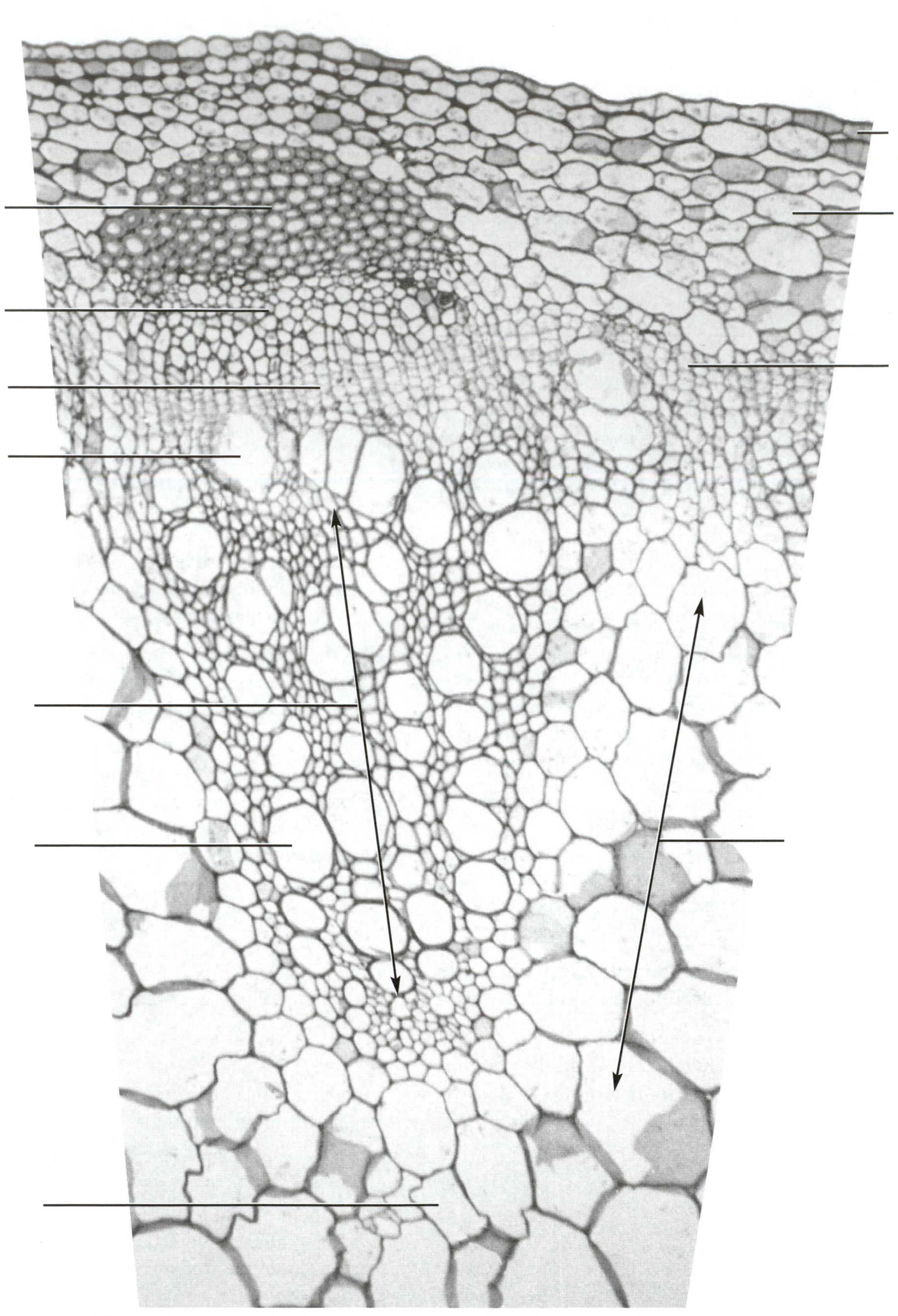

FIGURE 3.8 *Helianthus* stem, cross section of a vascular bundle, X 225.

Questions

1. What is meant by primary growth? (11)

   What is meant by secondary growth?

2. What are the three primary meristems? What tissue(s) develops from each? (12)
   a. —
   b. —
   c. —

3. How does the *distribution* of the vascular bundles differ in monocots and dicots?
   a. Monocots: (13)
   b. Dicots:

4. What *type* of tissue comprises the bundle cap of *Helianthus*? (14)

5. What are some differences between the xylem and the phloem? (15)

	xylem	phloem
relative sizes of the tissues		
relative thicknesses of the cell walls		
kinds of cells present		
sizes of the cells		

6. Briefly explain four functions of stems. (16)

   a.

   b.

   c.

   d.

Stems: Woody Plants

Introductory Notes

In Exercise 3, the stems of herbaceous plants were examined. Parenchyma tissue is the most abundant type of tissue found in herbaceous plants, which is why these plants remain relatively soft as long as they live. However, the stems of woody dicots develop hard tissues very soon after the plant begins growth.

Herbaceous plants are composed largely of primary tissues. While all stems have these primary tissues and increase in length by their formation, many stems also show a significant increase in diameter. This increase in diameter is caused by the formation of secondary tissues in the stem. These tissues develop through the activities of lateral meristems or cambiums. Two types of cambiums may be present: the vascular cambium and the cork cambium. Even though the emphasis in this exercise will be on an examination of woody stems and their secondary tissues, it should be remembered that there are also similarities with herbaceous stems.

GOALS

After completing this exercise, the student should be able to:

- make a diagram showing the external features of a woody dicot twig, labeling the terminal bud, lateral bud, bud scales, terminal bud scale scars, leaf scars, vascular bundle scars, and lenticels.
- explain how and why the terminal bud scale scars can be used to determine the age of a woody twig and be able to demonstrate this.
- identify, from a photograph or diagram, the structures seen in a cross section of a woody twig. These are listed in the exercise.
- explain the difference between tracheids and vessel members; identify each from a photograph.
- differentiate between springwood and summerwood.
- tell the age of a twig by examining a cross section of one.
- list several differences between hardwood and softwood; identify each from a photograph.
- write a brief paragraph explaining the differences between radial, tangential, and transverse wood cuts; identify each from a photograph.
- differentiate between heartwood and sapwood.
- answer the assigned questions.

EXTERNAL FEATURES OF A WOODY DICOT STEM

Examine the demonstration of a longitudinal section of the terminal bud of a horse chestnut (*Aesculus hippocastanum*) twig. Note next year's young leaves and the small, conical stem tip which has structures similar to those seen in the *Coleus* stem tip examined in Exercise 3. This small, conical portion has leaf and bud primordia that will mature in two years.

Obtain a preserved horse chestnut twig and locate the following regions or structures.

1. The **terminal bud** is the large bud at the apex of the twig. It is covered with modified leaves called **bud scales**. What is their function? (1)
2. **Leaf scars** can be seen as somewhat triangular regions along the stem and represent the place where the leaf petiole was attached to the stem. What is their arrangement? (If necessary, refer to Exercise 3.) (2)
3. Use a dissecting microscope to examine several leaf scars. The small dots within each leaf scar are **vascular bundle scars** and represent places where the xylem and phloem branched from the stem into the leaf petiole. What is the most common number of these found in a leaf scar? (3)
4. **Lateral** or **axillary buds** are found just above a leaf scar. (Remember the general relation of the leaf primordia and the bud primordia from Exercise 3?) Not all twigs will have a full complement of lateral buds, since these frequently get damaged during the growing season.
5. **Lenticels** are the small dots along the bark of the twig. These are regions where the cork tissue is not completely developed. What is their function? (4)
6. **Terminal bud scale scars** appear as rings of cork tissue encircling the stem. They mark locations of the

terminal bud in previous years. When the tissues inside the terminal bud begin to grow, the bud scales fall off, leaving these scars.

Make a diagram of an *Aesculus* twig that is *in its third year of growth*. (This means that you will have to modify your diagram from the actual age of your specimen). Label all of the seven previously mentioned structures (Fig. 4.1). Indicate the one-year old and two-year old portions of the twig.

FIGURE 4.1 Two-year-old *Aesculus* twig.

INTERNAL ANATOMY OF A WOODY TWIG

Examine a prepared slide of a cross section of a basswood (*Tilia americana*) stem. The slide has sections from one-, two- and three-year-old stems. Examine the one-year- old section first. The tissues are listed as they appear from the center of the stem toward the exterior.

1. Locate the central **pith**. This is surrounded by a ring of **primary xylem.**
2. The cylinder of tissue next to the primary xylem is the **secondary xylem.**
3. The **vascular cambium** can be seen as a narrow cylinder around the outside of the secondary xylem. Usually, this will appear as a curved line of darkly stained cells.
4. The **secondary phloem** is next in order from the center to the outside. This is composed of **phloem fibers, sieve tubes,** and **parenchyma.** (Companion cells are also present but difficult to identify.)
5. The **primary phloem** is at the outermost part of the phloem tissue. It may be difficult to distinguish this from the secondary phloem.
6. The **cortex** lies just external to the phloem. At this young age, there may be some epidermal tissue still present also.

After you have become familiar with the anatomy of a one-year-old twig, examine the cross section of a three-year-old stem.

1. Locate the secondary xylem. Notice the **annual rings.** Each annual ring contains both **springwood** and **summerwood.** How do the cells in these two regions differ? (5)
2. The secondary xylem contains several kinds of cells: **large vessels,** which are the chief conducting cells of the xylem; **tracheids,** which are smaller in diameter and serve as both conducting and supporting cells; **fibers,** supporting cells with thick cell walls; and **parenchyma** cells.
3. The **vascular rays** are lines of parenchyma cells that radiate outward from the pith (Similar to the pith rays seen in the sunflower). As in the sunflower, these separate individual vascular bundles, however, the individual bundles are smaller and more numerous. The rays are also smaller; some are only a single cell wide. Notice that some of these rays become dilated (expand) when they reach the phloem and appear as V-shaped masses of parenchyma. Rays are referred to as **xylem rays** (or **wood rays**) if they are in the secondary xylem, and as **phloem rays** if they are to the outside the cambium.
4. This older section should have layers of **cork** replacing the epidermis which has been lost as the stem increased in diameter. This tissue develops from another lateral meristem, the **cork cambium**
5. All of the tissues outside the vascular cambium collectively comprise the **bark** of a tree.
6. Figure 4.2 is a photograph of a four-year-old *Tilia* stem cross section. Label the **pith, primary xylem, secondary xylem, springwood, summerwood, annual ring, vascular cambium, secondary phloem, xylem ray, dilated** and **nondilated phloem ray, cortex,** and **cork.** Indicate the **first, second, third,** and **fourth** years of growth.

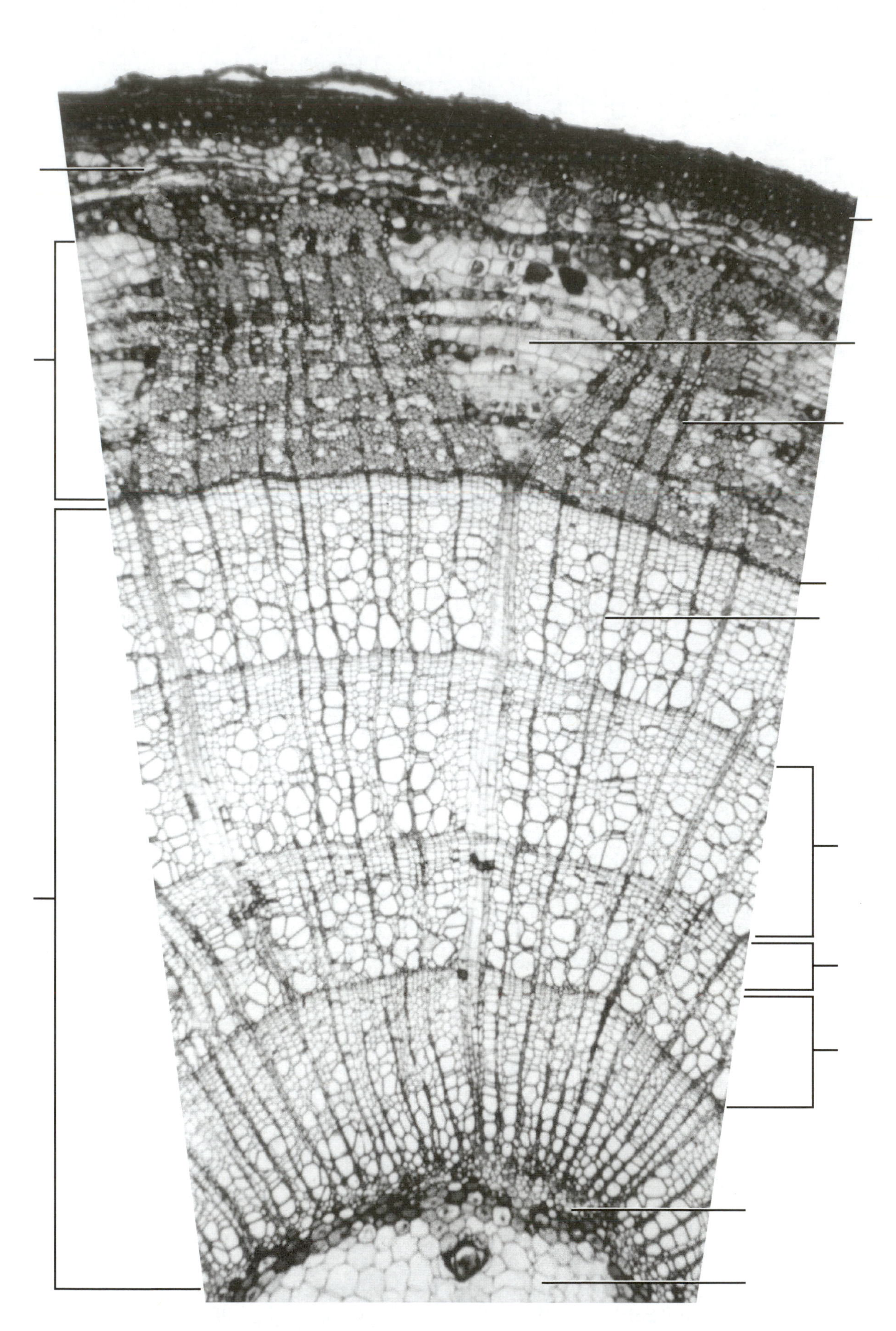

FIGURE 4.2 Cross section of a *Tilia* stem, X 120.

A MORE DETAILED STUDY OF THE SECONDARY XYLEM

Obtain a prepared slide of the macerated wood of *Magnolia* or other deciduous tree. Deciduous trees are commonly called hardwoods. The two main types of conducting cells can be seen in the following preparations.

1. Use low power and observe the two types of elongated cells. The larger cells are xylem **vessel elements**. Examine one of these in more detail, and notice the large perforations at the end of each cell, with the end wall nearly gone. There are only thin strands of cell wall material left, giving the appearance of a ladder. In other plants, this end wall may be completely absent. This allows water to flow freely from one cell to another. The small openings along the side of the cell are referred to as **pits**. These allow lateral movement of water.
2. The other, more slender, elongated cells are either tracheids or fibers. **Tracheids** are conducting cells, whereas the fibers do not conduct any liquid.
3. Diagram a single vessel element of *Magnolia*, labeling the **perforated end wall** and **lateral pits**. Then diagram a single tracheid, showing how these two cells differ in size and general shape (Fig. 4.3).

FIGURE 4.3 Vessel element and tracheid of *Magnolia* compared.

Now obtain a prepared slide of the macerated wood of pine (*Pinus*), a softwood. Examine this preparation with low power. How does this tissue differ from that of the magnolia? (6)

1. The slender, elongated cells are **tracheids**. The lateral perforations have a generally circular appearance with a raised ridge and are referred to as **bordered pits**.
2. Make a diagram of a single pine tracheid showing and labeling the **bordered pits** (Fig. 4.4).

FIGURE 4.4 Pine tracheid.

Wood can be sectioned or sawed in three different ways. You have already examined a cross section or **transverse** cut, where the cut is made at right angles to the main axis of the stem. The other two cuts are **longitudinal** sections. A **radial** section is one that is cut along the radius of the stem (a line extending from the center of the stem to the edge). A **tangential** cut is one made at right angles to the radius.

In a radial section, the wood rays can be seen in side view and appear as flat rows or sheets of cells. In a tangential section, the wood rays can be seen in end view only.

1. Examine a prepared slide of *Tilia*, or other deciduous tree, that has all three sections present. Be able to identify each type of section, the various cell types, wood rays, springwood, and summerwood.
2. Next, examine a prepared slide of *Pinus,* which also has all three sections present. Be able to identify each of the three types of sections. Pay particular attention to the radial section, which will show good views of the bordered pits on the tracheids.
   a. How do the rays of *Pinus* differ in size from *Tilia* or other deciduous wood? Use the tangential section to answer this. (7)
   b. Find the resin ducts in all three sections of *Pinus*.
   c. Figures 4.5 and 4.6 are photographs of transverse, radial, and tangential sections of maple (*Acer*) and pine (*Pinus*). They also illustrate some of the differences between hardwoods and softwoods.

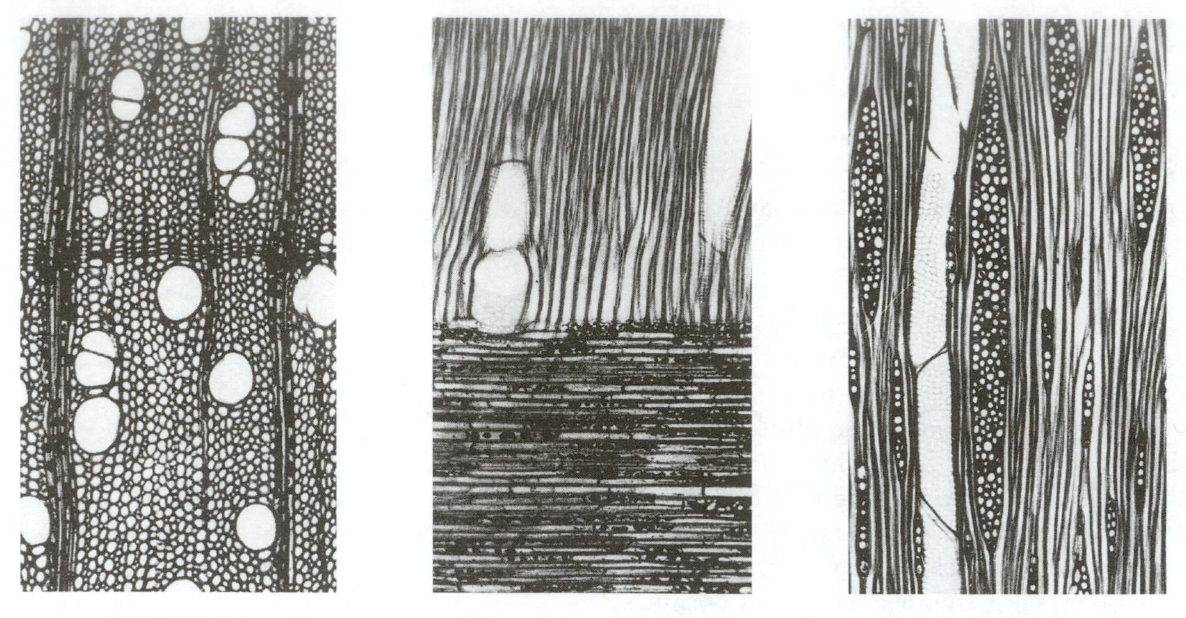

FIGURE 4.5 Transverse, radial, and tangential sections of *Acer saccharum,* X 75.

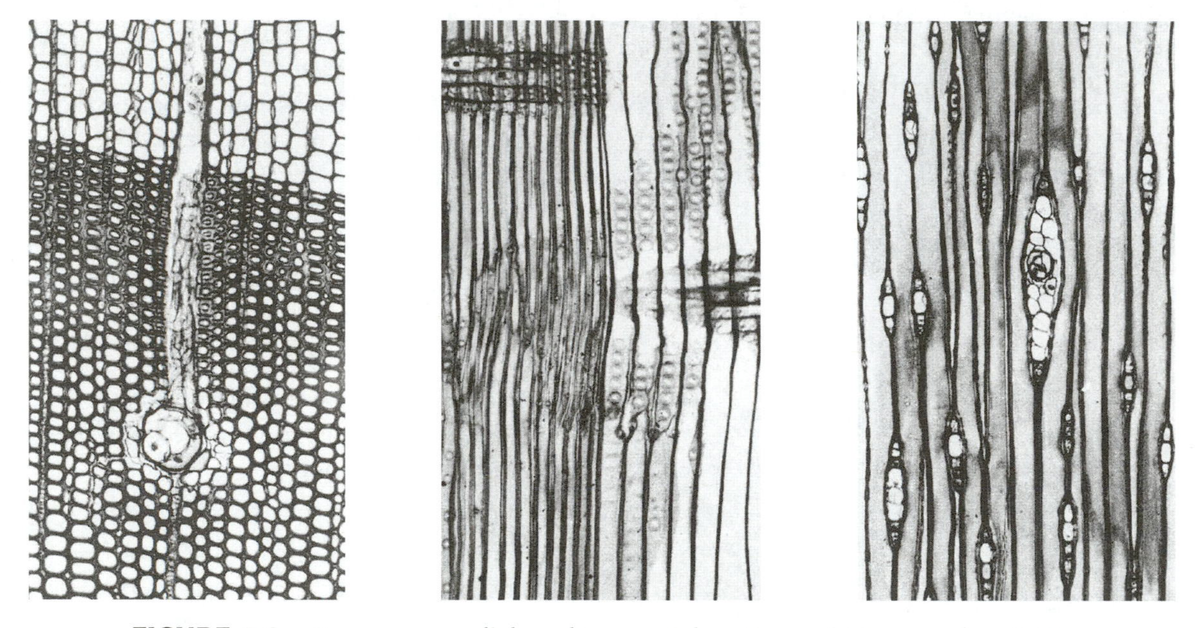

FIGURE 4.6 Transverse, radial, and tangential sections of *Pinus strobus,* X 60.

3. Thin slices of wood showing these three types of cuts from several tree species are available for you to observe these characteristics. Notice the type of grain in these species (caused by a combination of annual ring and wood ray features). Some of these are quite decorative and used for furniture or cabinets or as veneer.

4. Examine the larger blocks of wood on demonstration. You should be able to do the following:

   a. Identify the three types of wood cuts.

   b. Determine the age of each specimen.

Questions

1. What is heartwood? (8)

   What is sapwood?

2. What is the most abundant tissue in the 3-year-old stem? (9)

3. What are three **contrasting differences** in structure between the sunflower stem and the *Tilia* stem? (10)

Sunflower	*Tilia*
a.	a.
b.	b.
c.	c.

4. What are two techniques that can be used to determine the age of a woody twig? (11)

5. What causes growth rings in a tree? (12)

6. List four **contrasting features** that can be used to distinguish *Pinus* wood from *Tilia* wood? (13)

Pinus	*Tilia*
a.	a.
b.	b.
c.	c.
d.	d.

Roots

Introductory Notes

The root is the first organ to develop when a seed germinates. It is typically an underground structure that functions primarily in absorption of water and soil nutrients, conduction of these substances, anchorage of the plant and serving as food storage organs. Some plants have roots especially modified for this purpose, i.e., carrots, beets (both common and sugar), turnips, sweet potatoes, cassava, parsnips, and radishes.

EXTERNAL MORPHOLOGY OF ROOTS

Root Systems

1. Examine the demonstrations of a **taproot system** and a **fibrous root system.**
   a. How do these two differ in appearance? (1)

   b. How do they differ in their development? (2)

2. Examine the demonstration of corn prop roots. Roots such as these, which develop from stems or leaves, are called **adventitious roots.** Notice that these roots are growing from nodes at the base of the stem.
3. Make labeled diagrams of these three root systems (Figs. 5.1–5.3). Where appropriate, label the **primary root, secondary root,** an **adventitious root,** a **prop root,** and a **node.**

GOALS

After completing this exercise, the student should be able to:

- differentiate between: taproot systems and fibrous root systems, primary roots and secondary roots, and secondary roots and adventitious roots.
- identify each of the four growth regions of a root tip, and explain the main function of each region.
- identify the main tissues found in both a typical dicot root and in a typical monocot root from a photograph, as seen in cross section.
- list the main functions of these tissues.
- explain several differences between a dicot root and a monocot root.
- write a brief paragraph explaining how a branch root develops.
- compare and contrast the structures of roots with stems.
- answer the questions in the exercise.

FIGURE 5.1 Taproot system.

FIGURE 5.2 Fibrous root system.

FIGURE 5.3 Prop roots.

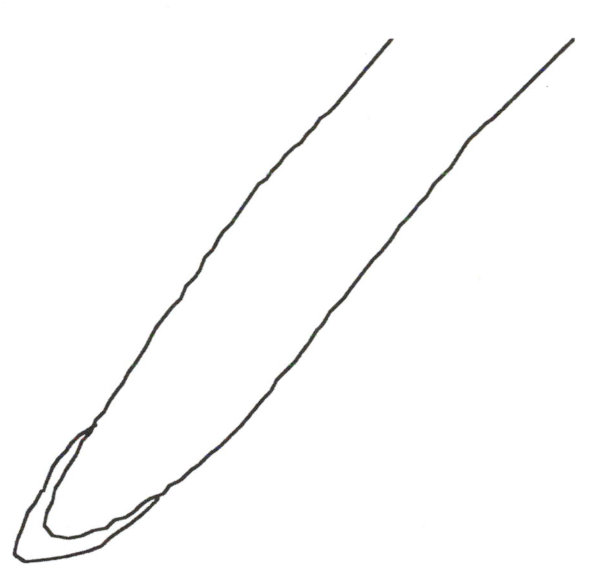

FIGURE 5.4 Longitudinal section of a root tip.

THE ROOT TIP

Regions of Growth

Examine a prepared slide of a longitudinal section of a corn root tip. Identify the following growth regions:

1. The **root cap** is a thimble-shaped mass of cells at the apex of the root, which protects the apical meristem from mechanical injury as the root grows through the soil.
2. The **meristematic region** is the region of active cell division. These cells should appear smaller and the region more dense than the cells in other regions.
3. Next is the **region of elongation.** Here, the cells produced by the apical meristem enlarge. Most of this size increase is along the main axis of the root, with the cells becoming decidedly more rectangular. The result of this activity is an increase in the length of the root (primary growth).
4. The **region of maturation** is where the cells begin to differentiate into their mature cell type. A darker, central portion of the root may be visible. This is the **stele** or vascular cylinder. Here you should be able to find some elongated vessel elements. The outer layer of cells is the **epidermis,** the cells of which should have slender projections, the **root hairs.** However, root hairs on prepared slides are often difficult to see clearly. The following exercise is designed to give you a greater appreciation of these structures.
5. Complete the diagram of a root tip (Fig. 5.4) by adding **root hairs** to show their general distribution. Label these plus *each* of the four different growth regions.

Root Hairs

Newly germinated radish seeds will be used to show root hairs. Before you make your preparation, notice the general appearance of a young root with root hairs and the huge increase in surface area these afford the root.

1. Make a wet mount of the root of a young radish seedling. Use forceps to transfer the root from the tray to your slide, and keep the root tip wet because the root hairs are very delicate. Remove the seed portion before adding your cover slip. After putting the cover slip in place, check to see whether the root is surrounded by water. If not, add a little from a dropper.
2. Examine the preparation with low power, paying particular attention to the youngest root hairs that will be found *closest to the root cap*. Note that the root hairs are extensions of the epidermal cells and not multicellular structures.
3. Add a drop of methylene blue dye to the edge of your cover slip, and draw it under by using a piece of paper towel at the opposite side of the cover slip to absorb water. The methylene blue should stain the nuclei of the cells. Look carefully to find one that shows the nucleus out in the root hair. This is what typically happens as the root hairs mature.
4. Make a labeled diagram of a single **epidermal cell** showing the attached **root hair** with the **nucleus** present in the root hair (Fig. 5.5)

FIGURE 5.5 Epidermal cell with root hair.

INTERNAL STRUCTURE OF ROOTS

Herbaceous Dicot Root

Examine a prepared slide of a cross section of a buttercup (*Ranunculus*) root. All of the tissues seen will be primary tissues. What does this mean? (3)

Using low power, note the three main regions of the root: the central vascular cylinder or **stele**, the **epidermis**, and the region between the epidermis and the stele, the **cortex**.

Examine the following tissues in more detail with high power:

1. The **epidermis** is a single layer of cells surrounding the entire root.
2. The root **cortex** of *Ranunculus* consists of several layers of cells, but these are not in an orderly pattern. Is this a simple or complex tissue? (4)

   a. What *type* of tissue is found here? (5)

   b. How does this compare with the cortex of the sunflower (*Helianthus*) stem? (6)

   c. Give a possible reason for this difference. (7)

3. Figure 5.6 shows the main regions of the *Ranunculus* root. Label the **epidermis, cortex,** and **stele.**

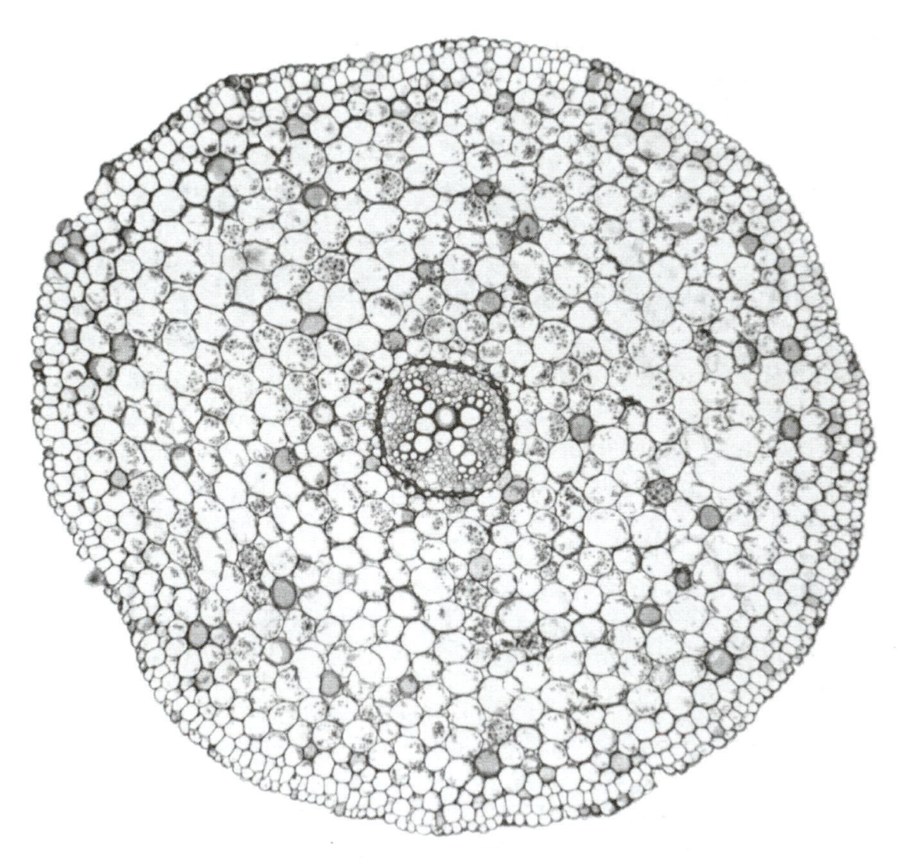

FIGURE 5.6 Cross section of a *Ranunculus* root, X 55.

The numerous, small structures in the cortex cells are **starch grains.** What does this tell you about the function of the root cortex? (8)

endodermis. The largest cells are the most recently formed vessel elements. Where are these located? (9)

4. The **endodermis** is the innermost layer of the cortex. The radial and transverse walls of the endodermal cells are impregnated with suberin, a waxy material. This band, the **Casparian strip,** restricts the movement of material into the stele.

   In the buttercup, some of the endodermal cells may have developed uniformly thickened cell walls. If so, try to find some that are not so thickened but still show the Casparian strip. These are referred to as **transfer cells.**

5. The **stele** includes all of the tissues enclosed by the endodermis.

6. The **primary xylem** occupies the very central portion of the stele. It is generally star-shaped or cross-shaped, with arms extending out toward the endodermis. The largest cells are the most recently formed vessel elements. Where are these located?

7. The **primary phloem** appears as small clusters of cells between the arms of the xylem. The phloem cells are much smaller than those of the xylem and also lack the secondary wall thickenings.

   Between the xylem and the phloem is a band of parenchyma cells. This is where the vascular cambium will develop in those plants that develop secondary tissues.

8. The **pericycle** is the single layer of cells next to the endodermis, between the endodermis and the primary vascular tissue. This tissue gives rise to the lateral roots.

9. Figure 5.7 is a close-up of the stele. Label the **cortex, starch grains, endodermis, transfer cell, primary xylem, primary phloem,** and **pericycle.**

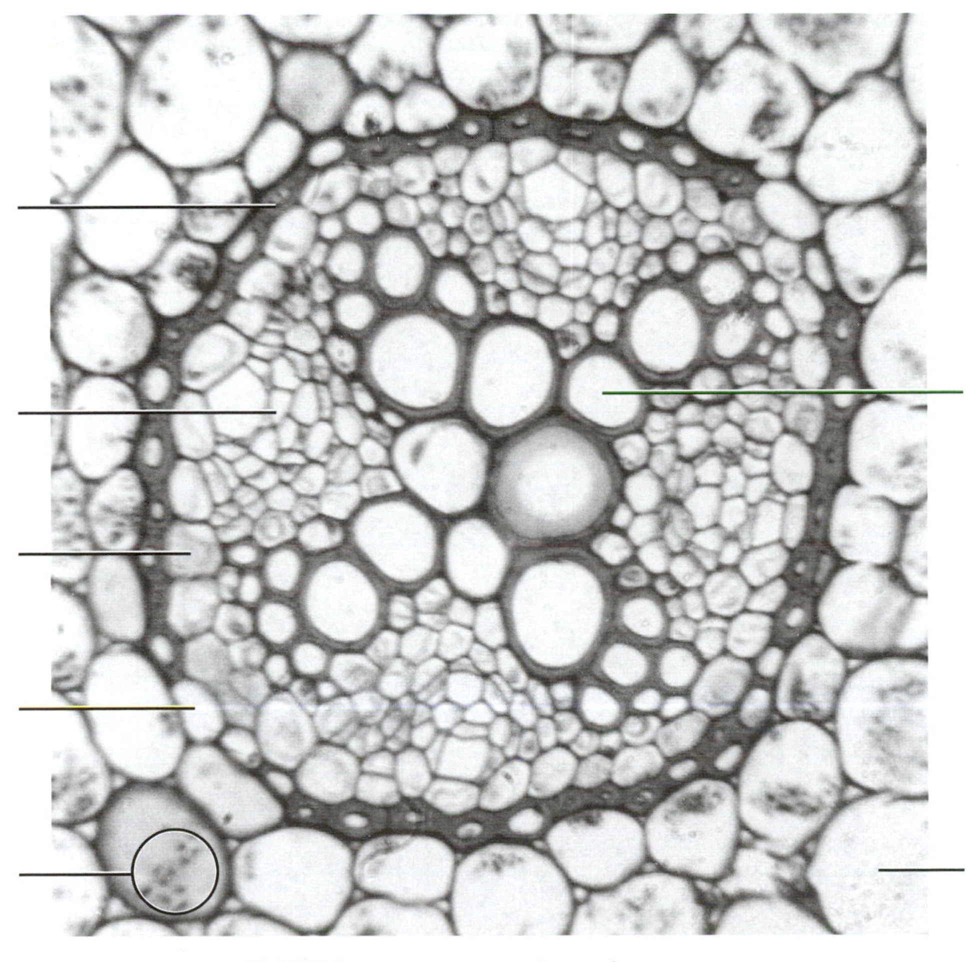

FIGURE 5.7 *Ranunculus* stele, X 250.

Monocot Root

Examine a prepared slide of a cross section of a *Smilax* root. Notice that the root is composed of the same three main regions as were seen in *Ranunculus*.

1. Compare the relative size of the steles of the two plants. (10)

   Ranunculus:

   stele = __________ the diameter of the entire root.

   Smilax:

   stele = __________ the diameter of the entire root.

2. Notice the layer of lignified cells beneath the epidermis. This is the **exodermis,** a tissue that develops in most roots.

3. The **endodermis** is very prominent and appears as U-shaped cells. Where are the cell walls the *least* thickened? (11)

   Smilax has a multiseriate **pericycle** (several cells thick), but it is found in the same relative location as in the *Ranunculus* root.

   a. What tissue occupies the central portion of the *Smilax* stele? (12)

   b. How does this compare to *Ranunculus*? (13)

4. The **primary xylem** and **primary phloem** alternate with each other, just inside the pericycle. Of these, the primary xylem is the more conspicuous.

5. Label the **epidermis, exodermis, cortex, endodermis, pericycle, primary phloem, primary xylem,** and **pith** of a *Smilax* root (Fig. 5.8).

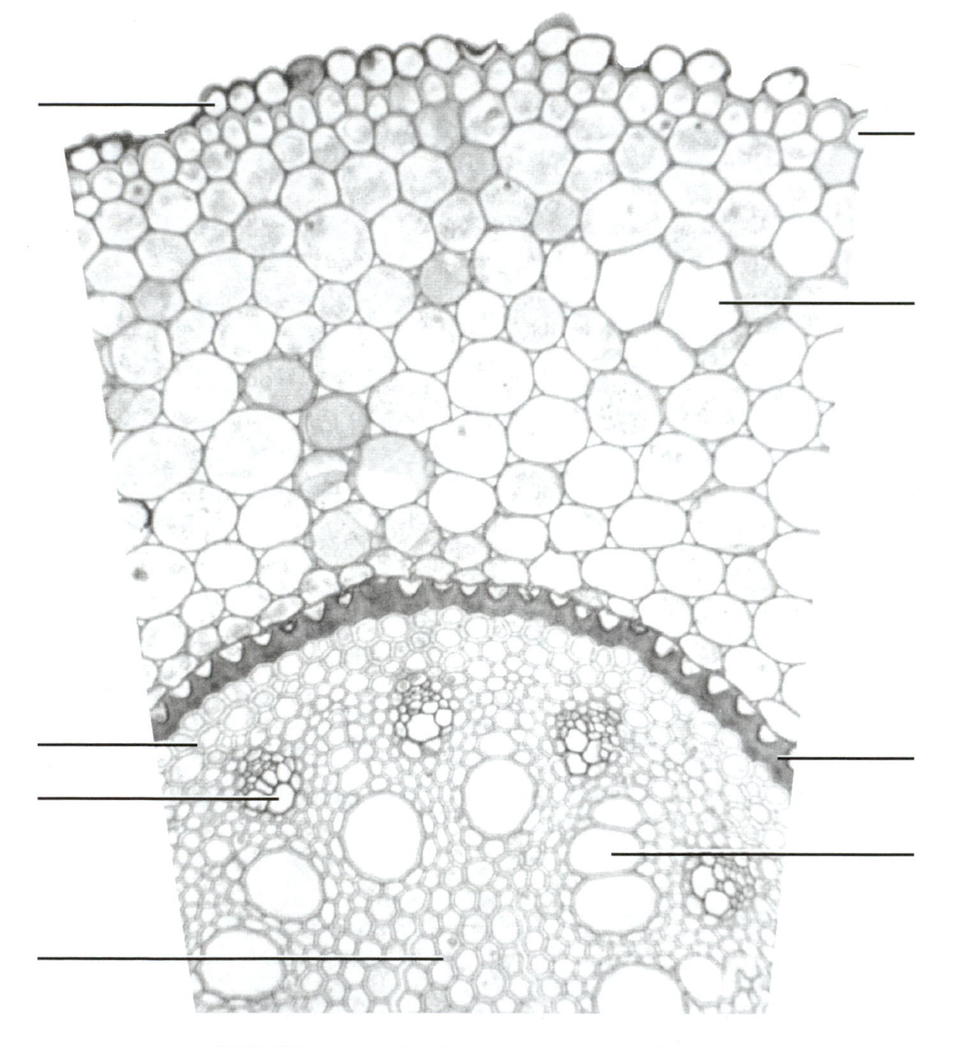

FIGURE 5.8 *Smilax* root cross section, X 170.

Woody Dicot Root

Examine a prepared slide of *Tilia* root cross section. *Tilia* and *Ranunculus* are both dicots, and their roots are similar in one important aspect. What might this be? (14)

1. Which tissue comprises the bulk of the *Tilia* root? (15)

2. Label the photograph of a *Tilia* root cross section in Figure 5.9.
3. Use your knowledge of *Tilia* stem anatomy and of the *Ranunculus* root to find and label the **annual growth rings, primary xylem, secondary xylem, vascular cambium, secondary phloem, bark, wood ray, phloem ray, cork,** and a **vessel element.**
4. Indicate the ages of the two annual rings indicated.

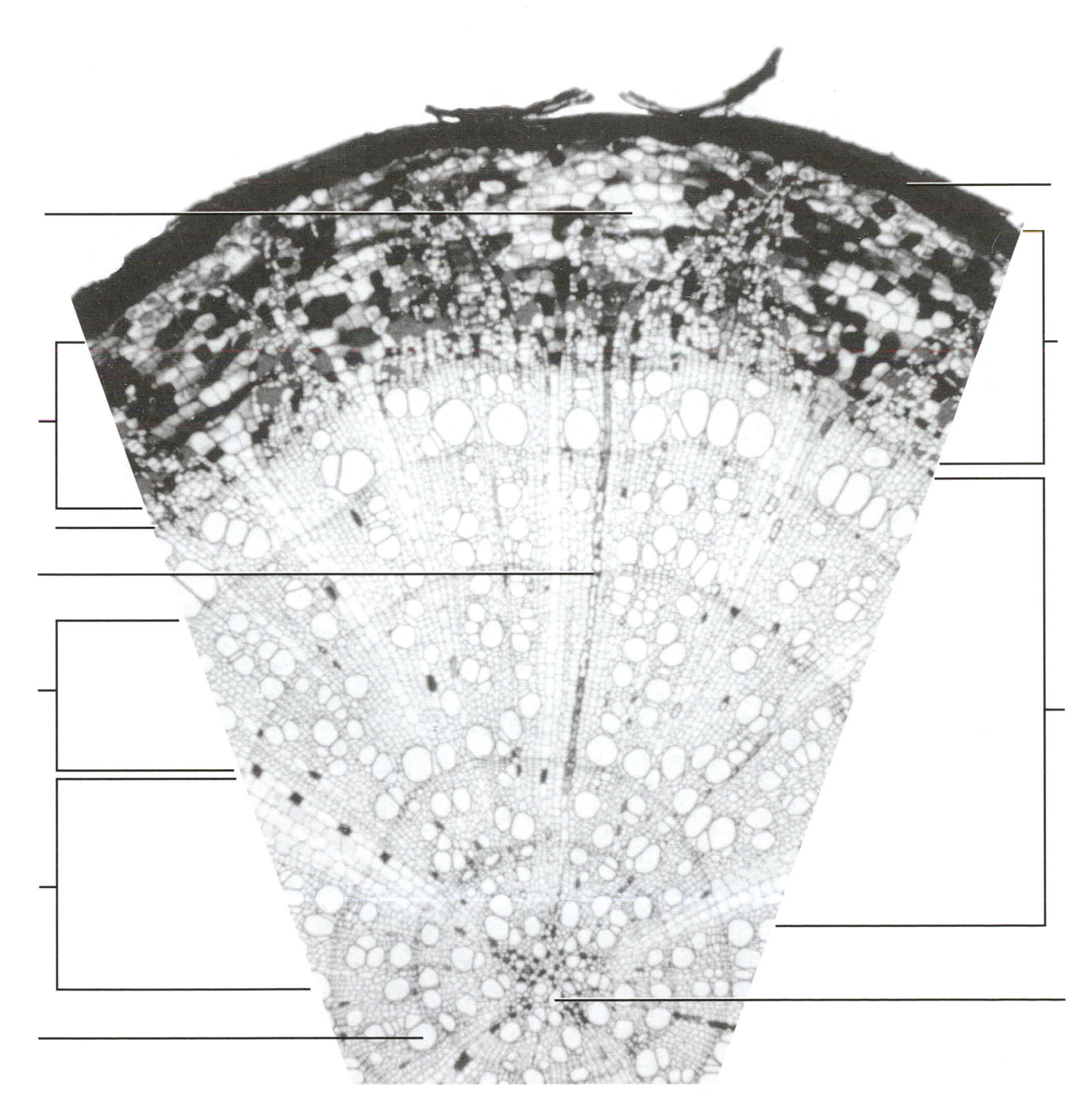

FIGURE 5.9 *Tilia* root cross section, X 120.

Lateral Root Origin

Obtain a prepared slide of a *cross section* of willow (*Salix*) roots showing different stages of branch root development.

1. Observe the section that shows the first stages of development.
2. Does the lateral root develop adjacent to the primary xylem or the primary phloem? (16)
3. From what tissue does the lateral root develop? (17)
4. Try to locate the root cap at the tip of a newly formed root.
5. Examine a prepared slide of a *longitudinal section* of a willow root showing the development of secondary roots. Be able to locate the **cortex, vascular tissues**, and, if possible, the **pericycle**.

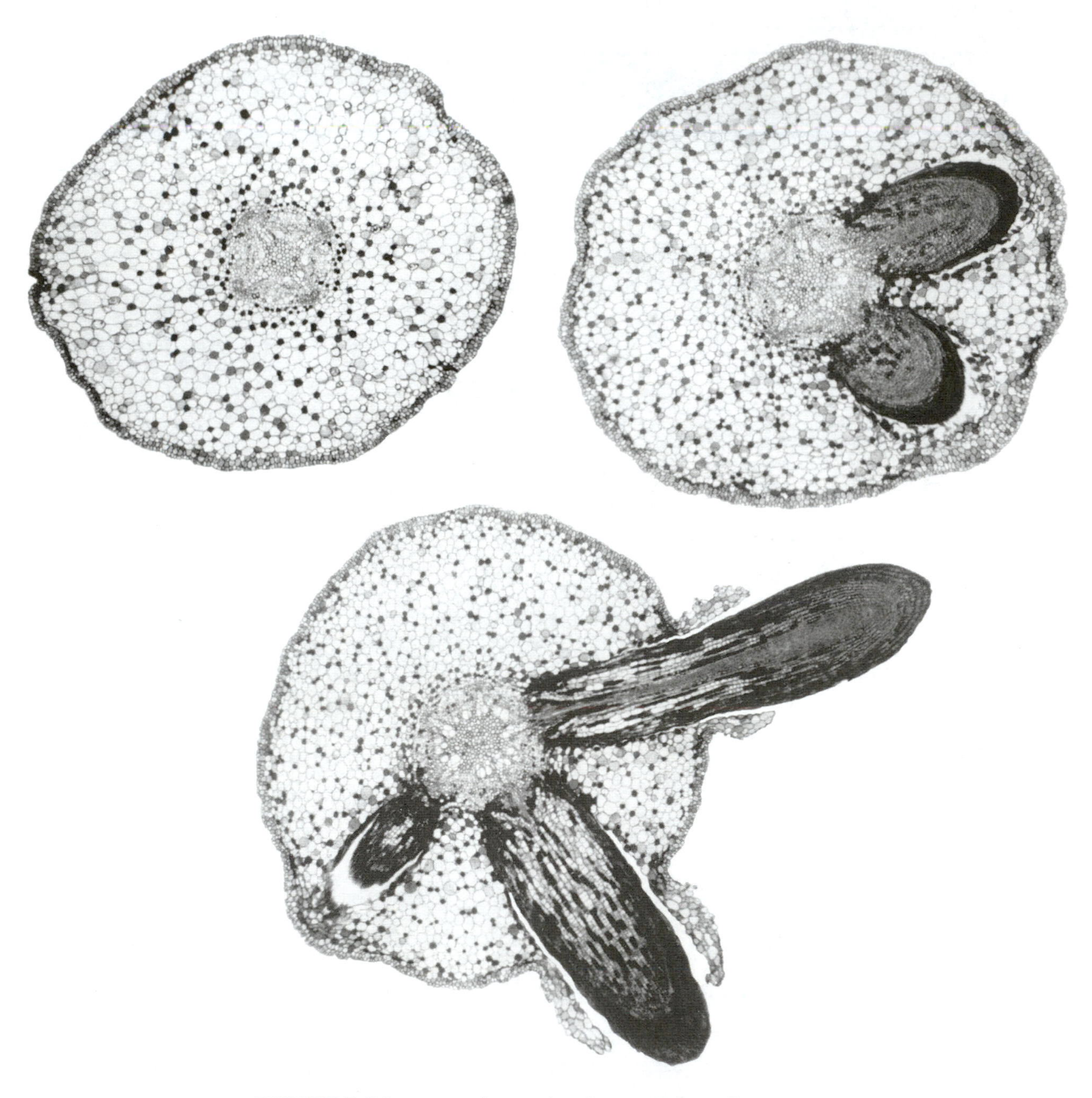

FIGURE 5.10 Lateral root development in *Salix,* X 120.

Questions

1. What is the **function** of the root hairs? (18)

2. What do root hairs do to the root to make them carry out their function more efficiently? (19)

3. What is the function of the pericycle? (20)

4. What is the function of the root cortex? (21)

5. What would happen to the root hairs if they were formed immediately behind the apical meristem instead of in the zone of maturation? (22)

6. List four contrasting features of monocot roots and dicot roots as seen in cross section. Use *Ranunculus* and *Smilax* as examples. (23)

Monocots	Dicots
a.	a.
b.	b.
c.	c.
d.	d.

7. List three contrasting features between dicot roots and dicot stems as seen in cross section. Use the *Ranunculus* root and the *Helianthus* stem as examples. (24)

Roots	Stems
a.	a.
b.	b.
c.	c.

8. List three contrasting features of the *Tilia* root and the *Tilia* stem as seen in cross section. (25)

Root	Stem
a.	a.
b.	b.
c.	c.

9. List two contrasting features of a monocot root and a monocot stem as seen in cross section. Use the corn stem (*Zea*) and the *Smilax* root as examples. (26)

Root	Stem
a.	a.
b.	b.

10. Compare the development of a lateral root with the development of a lateral branch. (27)

A lateral root

A lateral branch

Leaves

Introductory Notes

Leaves are the principle photosynthetic organs of green plants. The leaves of plants show considerable variation of form, size, structure, and arrangement on the twig. They are, however, well adapted, efficient food producing structures.

EXTERNAL MORPHOLOGY OF LEAVES

Examine the leaves of the florist's geranium (*Pelargonium*). This leaf consists of a broad, flattened **blade** or **lamina**, a leaf stalk or **petiole**, and a pair of small leaflike structures (**stipules**) found at the base of the petiole where it attaches to the stem.

A leaf having all three of these structures is said to be **complete**. If one or more of these is lacking, the leaf is **incomplete**. Usually, the stipules are absent, although they may be present in some modified form, such as thorns. Some leaves lack a petiole; the blade attaches directly to the stem. Such leaves are said to be **sessile**. Some leaves even lack a blade, i.e., tendrils of garden peas.

GOALS

After completing this exercise, the student should be able to:

- identify the main parts of a leaf.
- differentiate between complete and incomplete leaves.
- differentiate between simple and compound leaves.
- identify the various types and parts of compound leaves.
- explain the patterns of venation found in dicots, and contrast these with the typical pattern found in monocots.
- identify the internal parts of both monocot and dicot leaves.
- explain the difference between mesophytes, hydrophytes, and xerophytes and be able to recognize each of these from a photograph.
- describe the events of leaf abscission, and identify the parts of the abscission zone from a photograph.
- answer the questions in the exercise.

Make a diagram of a single *Pelargonium* leaf attached to the stem. Label the **blade, petiole, stipules**, and **stem** (Fig. 6.1).

FIGURE 6.1 *Pelargonium* leaf.

LEAF VENATION

You have compared monocot and dicot roots and stems. They also have different leaf structures. One major and easily seen feature is the pattern of the veins or vascular bundles in the leaf (the venation).

Monocots characteristically have a venation pattern where the veins are more or less **parallel** to each other, running the length of the leaf. These leaves have relatively few cross veins connecting these main vascular bundles.

Make a diagram of the monocot leaf on demonstration, showing parallel venation (Fig. 6.2).

FIGURE 6.2 Monocot leaf.

Dicots have **netted venation.** The veins branch many times in a random manner to form a network of smaller veins in the leaf. Leaves showing netted venation are on demonstration. Two variations of this pattern may be found:

1. **Pinnate venation.** Here, there is one main vein or midrib extending from the petiole through the blade with lateral veins branching from this at various intervals.
2. **Palmate venation.** Leaves with this pattern lack the single midrib. Instead, they have several main veins coming from one common point near the base of the blade.

TYPES OF LEAVES

Leaves may be categorized by the condition of the blade. If the leaf has a single, intact blade, it is said to be a **simple** leaf. If the leaf blade is so indented as to be separated into several smaller units, the plant has a **compound** leaf. Each of the individual units of a compound leaf blade is referred to as a **leaflet**. Some leaves have the margin undulated or indented somewhat, but if this undulation does not separate the blade into separate parts, it is a simple leaf.

Compound leaves follow the same venation pattern as simple leaves. **Palmately compound** leaves have all the leaflets arising from a common point at the end of the petiole. (In many cases, the individual leaflets may have a pinnate pattern.) **Pinnately compound** leaves have their leaflets arising at intervals along a main axis or **rachis**. This rachis is an extension of the petiole into the leaf blade. In addition, there are two types of compound leaves: those with an even number of leaflets and those with an odd number.

Examine the herbarium mounts of some common trees, and determine the leaf type and venation pattern of each (**simple/compound, palmate/pinnate**). Diagram a single, *intact*, leaf of each in Figures 6.3–6.6. Indicate if the *leaf* is simple or compound and whether the *leaf* has pinnate or palmate venation.

FIGURE 6.3 Beech (*Fagus*) leaf.

___________/___________

FIGURE 6.4 White ash (*Fraxinus*) leaf.

___________/___________

FIGURE 6.5 Maple (*Acer*) leaf.

___________/___________

FIGURE 6.6 Horsechestnut (*Aesculus*) leaf.

___________/___________

Internal Structure of Leaves

Dicot Leaf

Examine a prepared slide of a cross section of a lilac (*Syringa*) leaf. There are three major tissue types present: the **epidermis** (upper and lower); the **mesophyll**, or the tissue between the two epidermal layers; and the **veins**, the tissue within the mesophyll that conducts water and nutrients, and also supports the leaf blade.

1. Examine the epidermis and note that in *Syringa* this is only a single cell thick.

   Look for a **cuticle** on the upper epidermis. In *Syringa,* this may not be very prominent. The cuticle and epidermal cells function as a unit. What is their main function? (1)

2. Look for **guard cells** and associated **stomata** in the lower epidermis. This is a section of a leaf, so these cells will not have the characteristic bean-shaped appearance seen in Exercise 2.

3. The mesophyll is separated into two distinct regions. The **palisade mesophyll** is found on the top portion of the mesophyll, while the **spongy mesophyll** occupies the lower part.

   a. What *type* of tissue comprises the mesophyll? (2)

   b. The small, oval structures in the mesophyll cells are chloroplasts. Which layer has the greatest concentration of chloroplasts? (3)

   c. What is the main function of each mesophyll layer? (4)

   Palisade mesophyll

   Spongy mesophyll

   d. Compare the interior of the epidermal cells to the interior of the mesophyll cells. What conspicuous difference can you see? (5)

   Epidermis

   Mesophyll

4. Examine the midrib or main vein. Which tissue is on top, the xylem or the phloem? (6)

5. Examine the leaf/stem model to see why this is so.

6. Label the **palisade mesophyll, spongy mesophyll, air space, guard cell, stoma, upper epidermis, lower epidermis,** and **cuticle** as seen in a *Syringa* leaf cross section (Fig. 6.7).

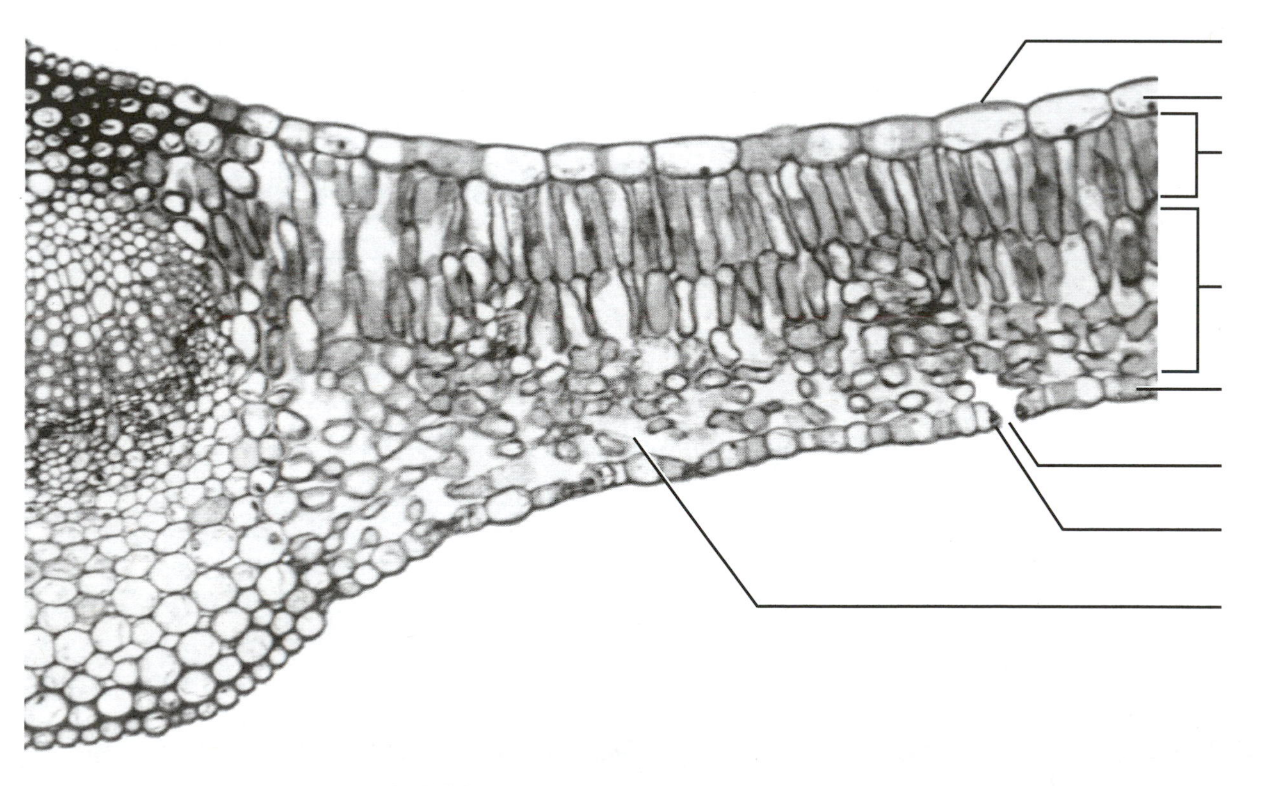

FIGURE 6.7 Cross section of a lilac (*Syringa*) leaf, X 170.

Monocot Leaf

Examine a prepared slide of a corn (*Zea mays*) leaf cross section. Notice that the mesophyll is not differentiated into two distinct layers, but consists of numerous parenchyma cells. Guard cells are found in both epidermal layers also.

The epidermis contains enlarged **bulliform cells.** It is thought that these cells lose water more rapidly than the surrounding epidermal cells. This causes the leaf to curl during dry conditions, possibly preventing excessive water loss.

Are these bulliform cells more abundant in the upper epidermis or in the lower epidermis? (Use information from question 6 to help determine this.) (7)

Examine the smaller vascular bundles. These will be surrounded by several large cells known as **bundle sheath cells.** These cells are very important to the photosynthetic process of corn.

The main vascular bundles resemble those of the stem (having the "face"). These bundles have conspicuous sclerenchyma fibers extending to each epidermis, called the **bundle sheath extension.**

Label the **upper** and **lower epidermis, bulliform cells, bundle sheath, primary phloem, primary xylem, mesophyll,** and **bundle sheath extension** of a corn leaf cross section (Fig. 6.8).

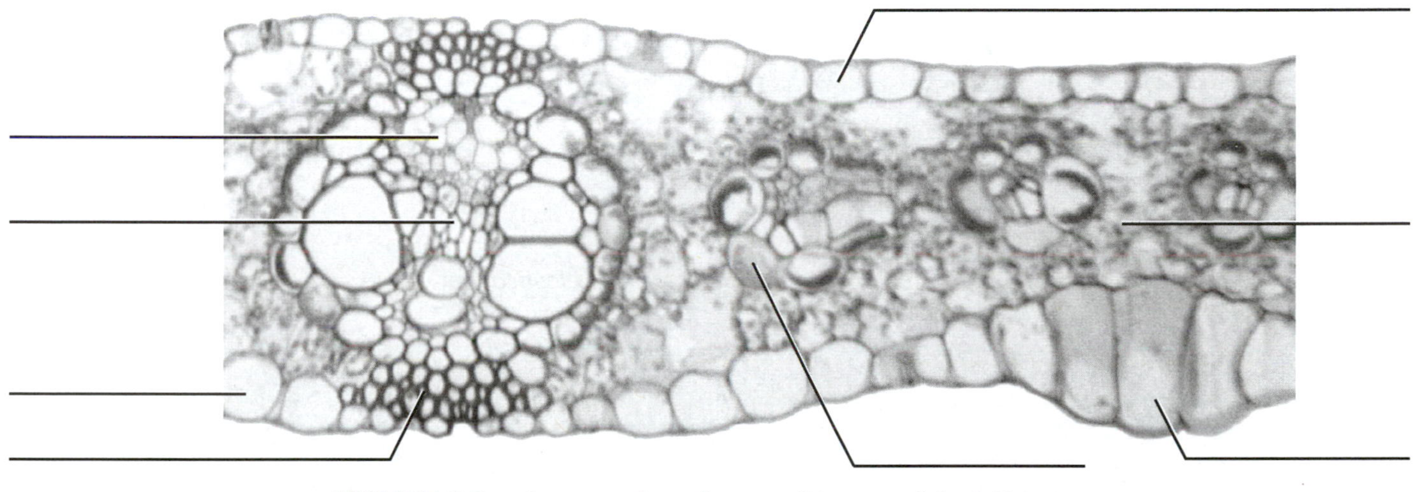

FIGURE 6.8 Cross section of a corn (*Zea mays*) leaf, X 200.

LEAF MODIFICATIONS

Plants, such as the lilac, that grow in environments that are neither extremely moist nor extremely dry are called **mesophytes.** Those plants that grow best in arid conditions are called **xerophytes,** while plants that grow in water are known as **hydrophytes.** Xerophytes have several adaptations for conserving water.

Examine a prepared slide of a *Ficus elastica* leaf cross section, and notice how it differs from *Syringa.* The top of the leaf has a very **thick cuticle,** below which are layers of clear cells. The epidermis has relatively small cells. The larger cells beneath the upper epidermis comprise the **hypodermis,** a water storing tissue.

The guard cells and stomata are located only in the lower epidermis and are sunken into small cavities (referred to as **stomatal crypts**). These are three typical features of xerophytes.

Now examine a prepared slide of a cross section of a water lily (*Nymphaea* sp.) leaf. Notice that there are many stomata located only on the upper epidermis. There is no cuticle, and the epidermis is composed of thin-walled cells. Throughout the mesophyll are star-shaped cells known as **astrosclereids.** Their function is not known.

What other adaptation can you see that indicates that this leaf floats on the surface of the water? (8)

Figure 6.9 is a photograph of a cross section of *Ficus elastica.* Label the **cuticle, upper epidermis, lower epidermis, hypodermis, palisade mesophyll, spongy mesophyll,** and **stomatal crypt.**

Label the **upper epidermis, lower epidermis, palisade mesophyll, spongy mesophyll, air chamber,** and **astrosclereid** of *Nymphaea,* (Fig. 6.10).

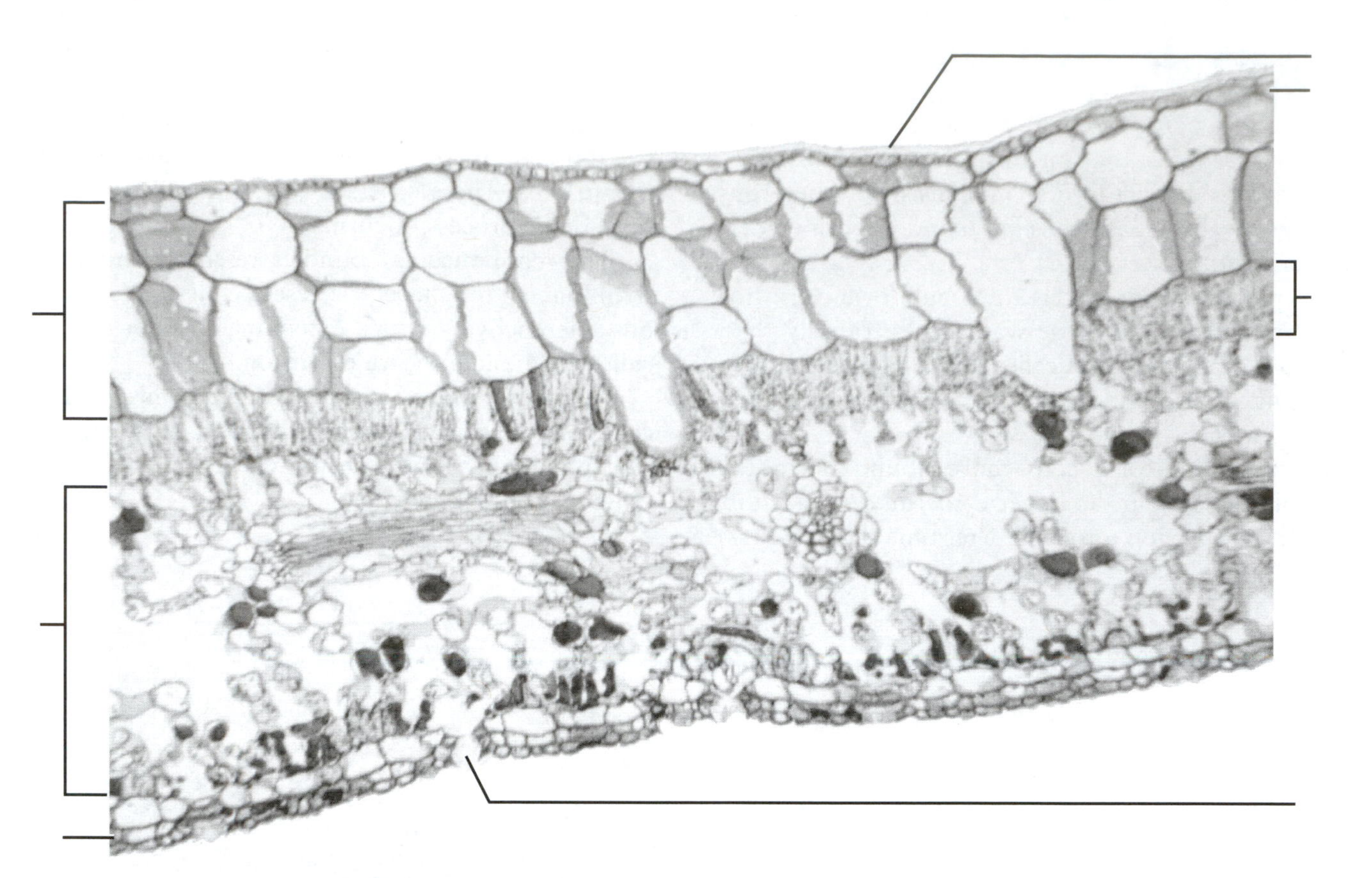

FIGURE 6.9 Cross section of a *Ficus elastica* leaf, X 145.

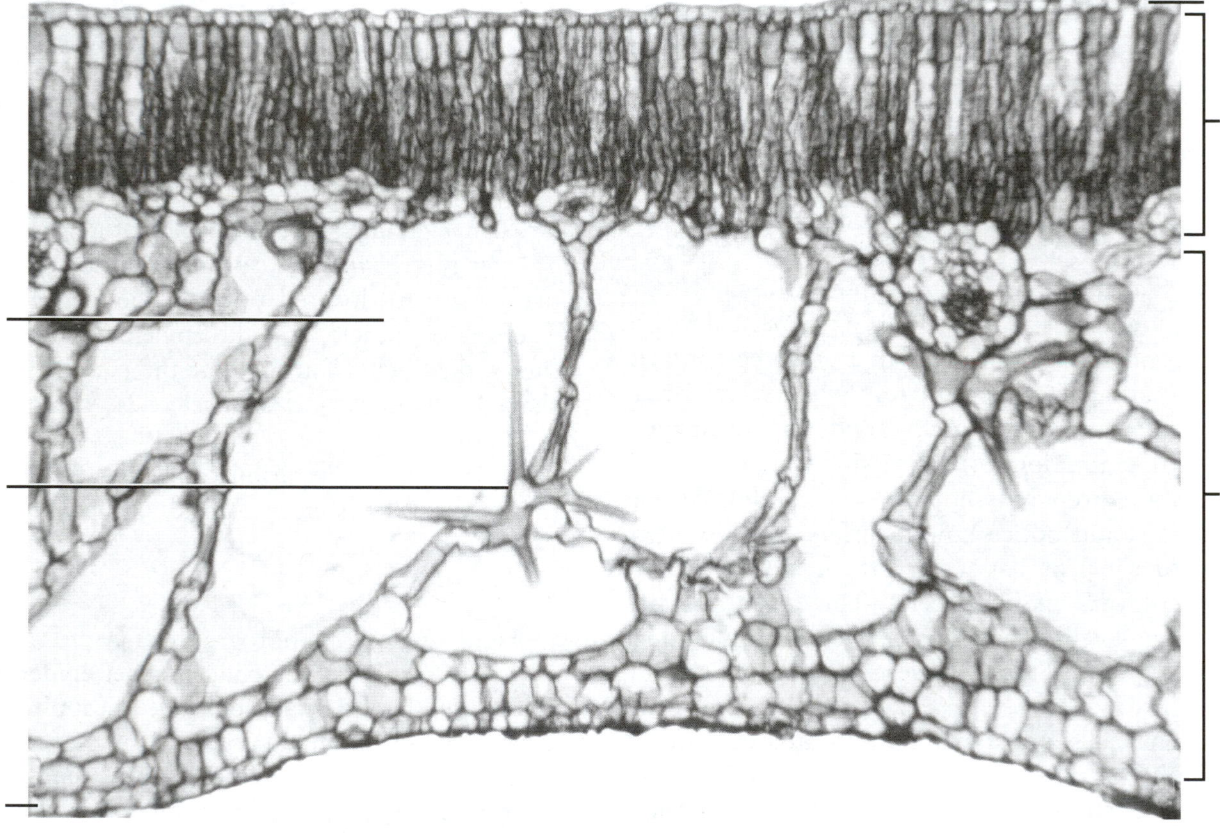

FIGURE 6.10 Cross section of a *Nymphaea* leaf, X 155.

Finally, examine a prepared slide of a *Nerium oleander* leaf cross section. Is this plant a hydrophyte, a mesophyte, or a xerophyte? (9)

What evidence did you use to draw your conclusion? (10)

Among the most interesting leaf adaptations are those of "carnivorous" plants. These include the well-known Venus's flytrap, the pitcher plant and the sundew.

A less-known, but more common plant with this adaptation is the bladderwort (*Utricularia*). This is a submerged, aquatic plant, commonly found in lakes. Some of the leaves of this plant form saclike **bladders** that have **sensitive hairs** that extend into the water near an opening of the bladder. When a small aquatic animal comes in contact with these hairs, the bladders quickly expand, drawing water and the small organism into the sac. Here they are slowly digested.

Examine a prepared slide of the bladderwort, and locate the small bladders and the hairlike "triggers." You may find captured organisms in some of the bladders.

Diagram a single bladder of *Utricularia*, showing and labeling the **opening** into the bladder, the **sensitive hairs**, and the **petiole** of the leaf (Fig. 6.11).

LEAF ABSCISSION

Most woody dicots in temperate regions produce new leaves each spring and lose them in the autumn. This loss of leaves is not a haphazard event, but the result of a series of changes and events. Leaf fall is associated with the development of a special layer of cells at the base of the petiole, the **abscission zone.** In this area are two types of cells in two layers. The **separation layer** has parenchyma cells that extend throughout the petiole except through the vascular tissue. Chemical changes brought on by changing photoperiod cause these cells to separate from each other (commonly by the breakdown of the middle lamella). At the same time, a **cork layer** develops, which protects the stem from excessive water loss. Eventually, the leaf is attached only by the vascular tissue, which forms corky plugs. Mechanical forces, such as wind, rain, etc., then cause the leaf to fall off.

Examine a prepared slide of a longitudinal section of a *Populus* twig that shows that abscission zone. You should be able to relate this to the leaf scars and vascular bundle scars that were seen on the woody twig. In the following photograph (Fig. 6.12), label the **leaf trace, leaf gap, cortex, pith, leaf petiole, vascular tissue, lateral bud,** developing **vascular bundle scar,** the **separation layer,** and **cork layer** of the abscission zone.

FIGURE 6.11 Bladderwort (*Utricularia*) leaf.

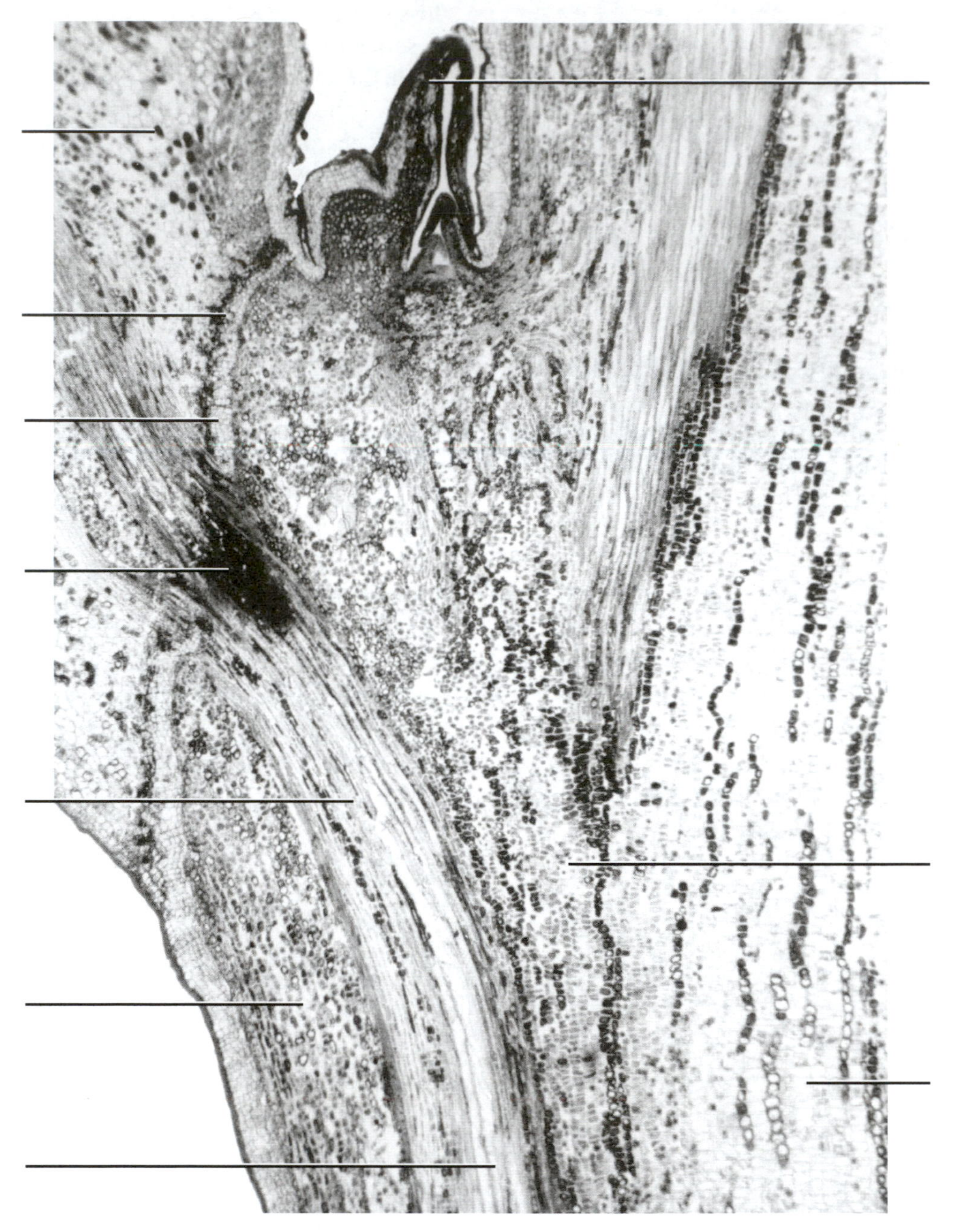

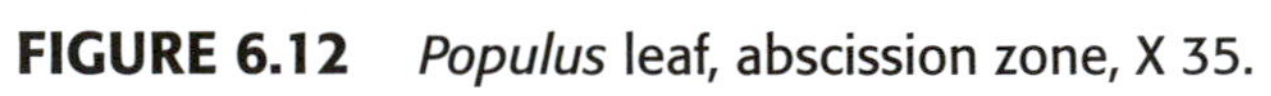

FIGURE 6.12 *Populus* leaf, abscission zone, X 35.

Questions

1. How can you tell whether a leafy structure is an entire leaf or merely a leaflet of a pinnately compound leaf? Assume the structure is intact and attached to a twig. (11)

   a. Leaf:

   b. Leaflet:

2. How does the mesophyll of a corn leaf differ from that of a lilac leaf? (12)

   a. Corn:

   b. Lilac:

3. What are the three special adaptations of xerophyte leaves? (13)

   1.

   2.

   3.

4. What are some features that you could use to distinguish a dicot from a monocot? List contrasting features for each of the following: (14)

	Dicot	Monocot
1. **Roots:** External Internal a. b.		
2. **Stems:** Internal a. b.		
3. **Leaves:** External Internal		

Some Aspects of Classification

The remaining exercises are given to a survey the major groups of plants. These are not intended to be a comprehensive study of these plant groups but will illustrate some of the important aspects of their structure, life histories, methods of reproduction, and the ecological importance of some of the more important plant groups.

It would be helpful if you read something about plant classification before starting these units. You should at least know that every plant (and all living things) belong to a category known as the species. For example, the common dandelion is of the species *Taraxicum officinale*. This scientific name is never used for any other organism. Each species then belongs to a category known as the genus, each genus to a family, each family to an order, each order to a class, and each class to a division. The division is the most inclusive category (similar to a phylum in zoology). If we again use the dandelion as an example, we have the following hierarchy of categories.

Division: Anthophyta (all flowering plants)
Class: Magnoliatae (all dicots)
Order: Asterales
Family: Asteraceae (the sunflower family)
Genus: *Taraxicum*
Species: *officinale**
(the common dandelion)

There is a fair amount of agreement among botanists about the more specific taxonomic levels, i.e., families, genera, and species. Unfortunately, the same cannot be said about the larger categories. Ideally, a classification system should show evolutionary trends and genetic relationships. To show these relationships between major taxonomic groups requires a great deal of information about plants that are long extinct. The fossil record is just not complete enough for us to determine these relationships with certainty.

This lack of information has led to the development of many hypotheses concerning plant evolution and to many systems of classification. Indeed, if one were to examine several introductory botany texts, it is very possible that several classification schemes will be proposed.

The taxonomic scheme presented in this manual is, with a minor alteration, that proposed by Bold.[1] Appendix A gives a comparison between Bold's system and one proposed by Tippo[2] a bit earlier. The most important difference is in the treatment of the vascular plants. Tippo placed all vascular plants in a single phylum. This implies that ferns and redwood trees are closely related. Bold, on the other hand, believed that the differences between the various groups of vascular plants are more significant than the single attribute of having xylem and phloem tissues. He, therefore, separated them into several divisions (hinting also that these plants probably did not have a common ancestor). Earlier systems of classification had as few as three divisions: algae and fungi, mosses and liverworts, and finally, all vascular plants. Hopefully, you will see some of the problems inherent in this approach as you study the various plant groups.

* More properly, this should be called the specific epithet or specific name. The species really has two components, the genus plus the specific name.

One other problem of classifying organisms that is worth mentioning here is the grouping of organisms into kingdoms (the most inclusive category). Both Bold and Tippo used only two kingdoms (plants and animals). For convenience, this traditional approach is used in this lab manual also. However, you should be aware that all organisms do not fall neatly into these two categories. For example, we will study a single-celled organism that contains chloroplasts and carries on photosynthesis — an obvious plant feature. This organism (*Euglena*) also swims through water quite well — an animal feature. Moreover, if it is placed in the dark for an extended period of time, the chlorophyll will deteriorate, and it may not be able to produce more when it is returned to the light. When this happens, *Euglena* must carry on a heterotrophic mode of existence. Finally, some organisms that are thought to be closely related to *Euglena* normally ingest solid food particles. Is *Euglena* a plant or an animal?

There are many other examples that could be used to show the difficulties encountered when trying to place all organisms into only two kingdoms. Because of this difficulty and because we have learned a great deal about the life histories and anatomy of the many species of living things, alternative classification schemes have been proposed. Whittaker[3] and Margulis[4] have given very convincing arguments for the establishments of five kingdoms. As might be expected, three and four kingdoms have been proposed. Many modern biology texts now have six kingdoms. One thing all of this does is point out that the whole idea of classification is arbitrary.

There is much more to the science of taxonomy than simply giving organisms hard to pronounce names and putting them into categories. A well-though-out classification scheme should show evolutionary relationships. Closely related organisms should be placed in the same category. The closer this relationship, the more specific the category should be. One other important aspect of classification involves the use of scientific names. This does away with problems inherent with the use of common names. Scientific names are never duplicated and are the same all over the world. This allows biologists to communicate with each other without having to know a variety of common names. For example, a common clubmoss that grows in the northeastern United States (*Lycopodium complanatum*) has 16 common names listed in one source.[5]

This lack of conformity at the kingdom level may not be due so much to the lack of information about the various species, but possibly to the unwillingness of biologists to accept new and different ideas. The various algae and fungi have traditionally been considered plants, but they are in four of the five kingdoms proposed by Whittaker. For most, the two kingdom approach is just simpler, even if at best they can claim to be studying plant-like organisms. Recent studies using DNA and gene analysis are further muddying the waters of the strange sea of classification.

Once again, the following exercises are not meant to be an in-depth study of plant classification or plant morphology. The main objective is to give the student a better appreciation of the tremendous diversity of plant life that exists. There is a great variety of structural complexity in the plant world, from unicellular algae to the world's largest living things — the giant sequoias of California.

1. Bold, H. 1957. *Morphology of Plants*. New York: Harper and Brothers, Publishers.
2. Tippo, O. 1942. A modern classification of plants. *Chronica Botanica* 7, 203–206.
3. Whittaker, R. H. 1969. New concepts of kingdoms of organisms. *Science* 163, 150–160.
4. Margulis, L. 1971. The origin of plant and animal cells. *American Scientist* 59, 230–235.
5. Brooks, K. L. 1979. *A Catskill Flora and Economic Botany: I. Pteridophyta, The Ferns and Fern Allies*. New York State Museum Bulletin No. 438. Albany. University of the State of New York, State Education Department.

The Plant Life Cycle

Regardless of any differences in appearance or structure, plants have certain aspects of their life histories in common. At some time an egg is fertilized, producing a diploid (2N) zygote. If a species is to maintain a constant number of chromosomes from one generation to the next, the process of meiosis must also occur at some time in the life cycle. This results in cells with the haploid (N) number of chromosomes.

In many algae and fungi, the only diploid stage is the zygote; the vegetative parts of the plant are all haploid. A few species of algae have life cycles similar to those of mammals, with the only haploid cells being the gametes. Some algae and fungi, and all the higher plants have a rather complicated life cycle involving what is known as **alternation of generations.** While the specific details of this process vary from one group of plants to another, the main features are similar in all groups.

As we study the various plant groups, frequent reference will be made to certain aspects of the life cycle of those plants. It is important that you understand the following essentials of alternation of generations:

1. The diploid zygote divides by **mitosis** to produce a mature plant, the cells of which are diploid in their chromosome content. This diploid plant is known as the **sporophyte** (spore producing plant). All of the diploid stage of the life cycle is known as the **sporophyte generation.**

2. Certain cells of the sporophyte, which are usually produced in sporangia, then divide by **meiosis**, producing haploid spores.

3. Typically, the haploid spores are released, germinate, and divide by **mitosis.** These mitotic divisions result in a haploid plant known as the **gametophyte** (gamete producing plant).

4. On the gametophyte plant, special structures develop that produce gametes (which are not always referred to as eggs and sperm). Since the cells of the gametophyte plant are already haploid, these gametes are produced by mitosis.

5. The egg is then fertilized, and a diploid zygote is produced, starting the next generation.

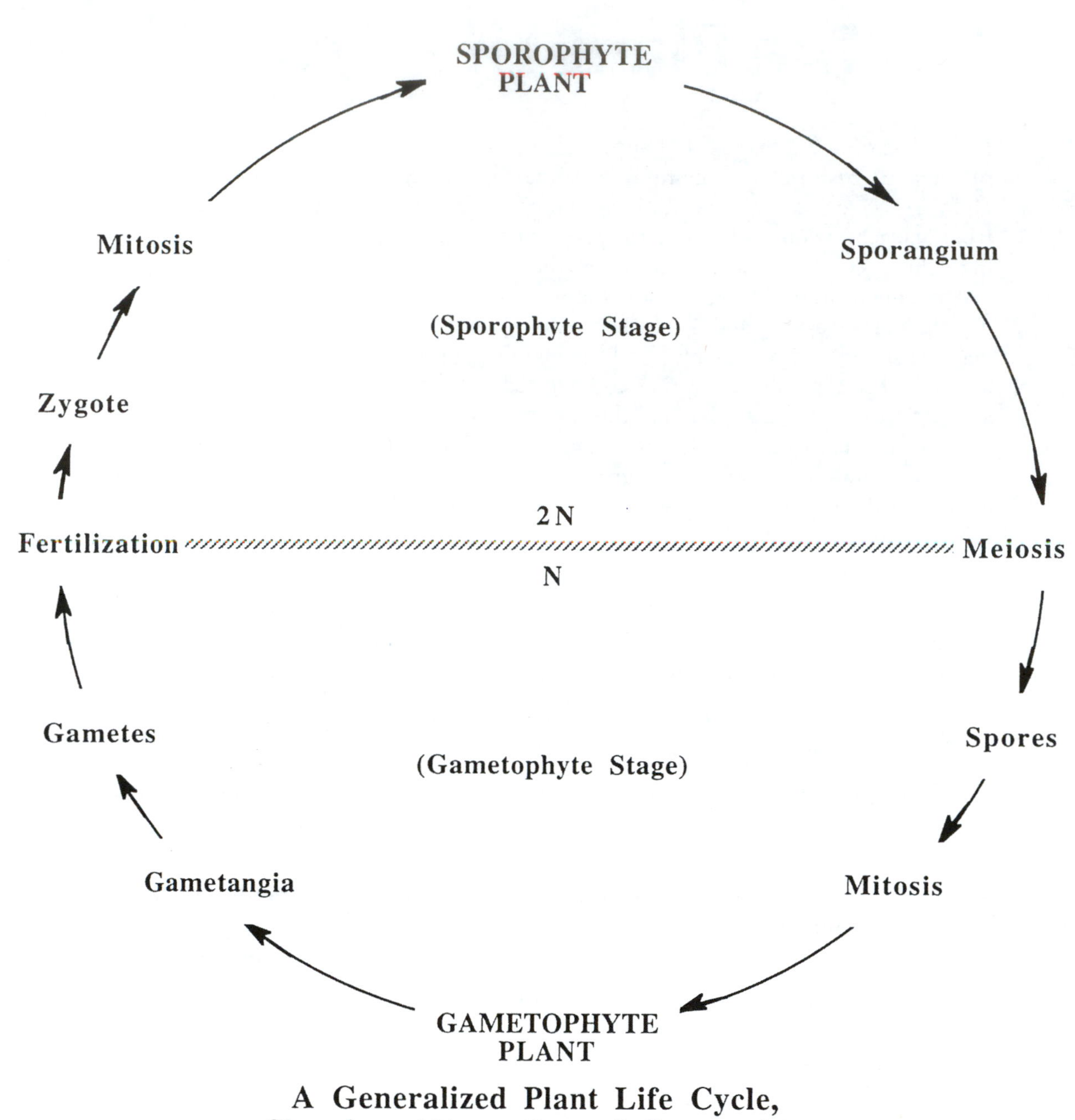

A Generalized Plant Life Cycle, Showing Alternation of Generations

Algae

Introductory Notes

An examination of some of the various groups of algae is a logical starting point for our examination of the plant kingdom. These organisms have a very simple structure, never producing true roots, stems, or leaves. They lack internal tissue differentiation, and the reproductive organs are commonly a single cell (if they are multicellular, then all of the cells are potentially fertile). Because of this relatively simple body plan, many biologists classify the algae as members of the kingdom Protista.

Most botanists also agree that the terrestrial plants originated from algal ancestors. Unfortunately, the fossil record is not complete enough for us to make too many definitive statements about these relationships. It should be emphasized, however, that present-day algae species are not ancestors to higher plants.

Algae are most commonly found in aquatic habitats, both freshwater and marine. In addition, many species may be found on tree trunks, in moist soil, and even on bare rocks. Most are microscopic organisms. However, this small size should not be equated with lack of importance. The algae are extremely important members of aquatic food chains and also contribute a substantial proportion of the oxygen in our atmosphere.

At one time, algae were classified as a single large group of plants. However, technological advances have provided us with considerable information about the anatomy, physiology, and methods of reproduction of algae. Now the term *algae* no longer implies a formal unit of classification. Instead, algae are separated into about ten divisions. These separations are based on (1) the photosynthetic pigments present, (2) the food reserve produced, (3) the chemical nature and composition of the cell wall, (4) the types of flagella, and (5) any special features of the cell structure. It is of interest to note that only one of these divisions, the Chlorophyta, has the same chlorophyll pigments, food reserve, and cell wall composition as the higher plants.

GOALS

After completing this exercise, the student should be able to:

- identify, from either photographs or diagrams, the representative genera of the divisions studied.
- explain the differences between the four growth forms seen (unicellular, colonial, filamentous, siphonous).
- differentiate between conjugation and oögamy.
- place the genera studied in their correct division.
- list the main characteristics of each division.
- answer the questions in the exercise.

DIVISION CHLOROPHYTA (GREEN ALGAE)

The great majority of green algae are found in fresh water. This is probably the most diverse and also the best studied of all the algal divisions. These studies have led to the identification of several evolutionary tendencies within the group. If pressed, many phycologists (individuals who specialize in the study of algae) would identify a single-celled organism of the group as the common ancestor of all the various evolutionary lines.

The important diagnostic features of this division are (1) both chlorophyll *a* and *b* are present, (2) starch is produced as the food reserve, (3) the cell walls always contain cellulose, and (4) if flagella are present, there are either two or four "whiplash" types.

Order Volvocales

The order Volvocales represent an evolutionary line of primitive green algae. Examine the representative genera in the following sequence, and note the increases in complexity.

Chlamydomonas (Fig. 7.1) is a single-celled, motile form that is considered the most primitive member of the order. *Chlamydomonas*, or a similar form, may be the evolutionary ancestor of all the green algae.

1. Make a wet mount of *Chlamydomonas*, find some that aren't moving about too much, and examine them with high power. Try to locate the two flagella at the anterior end. This will require careful adjustment of the light.
2. Look for the red **stigma**, generally found on the anterior third of the cell. This is a light sensitive structure.

Chlamydomonas has a single, cup-shaped chloroplast that takes up most of the cell volume.

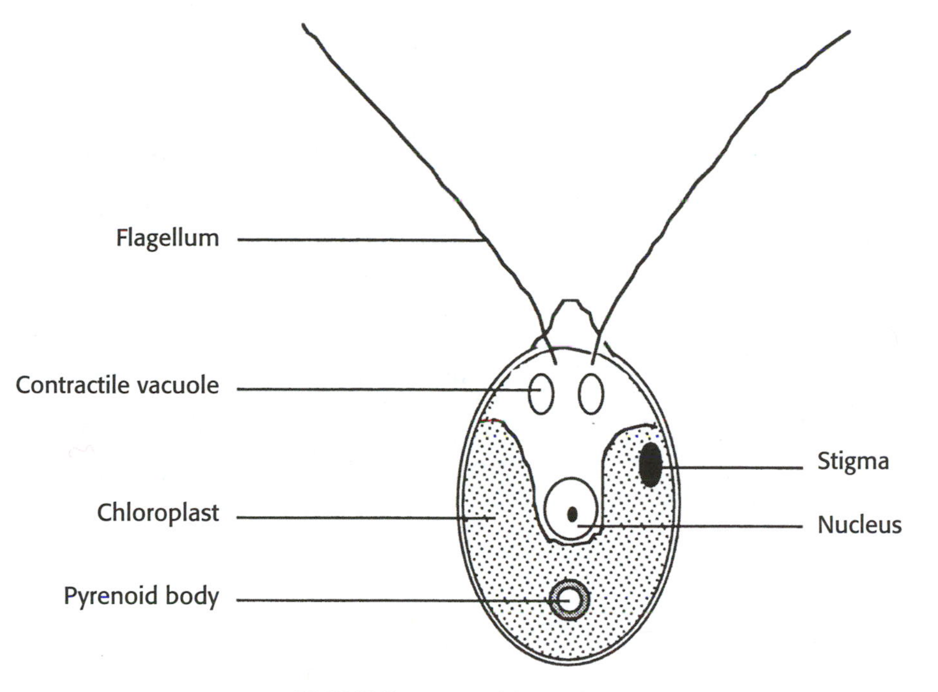

FIGURE 7.1 *Chlamydomonas,* X 2100.

Gonium is the simplest of the colonial algae in this order, consisting of 4 or 16 cells arranged in a flat plate and surrounded by a gelatinous envelope. Each cell in the *Gonium* colony, and in the others in the line, is similar to a *Chlamydomonas* cell. We will not examine *Gonium*.

Pandorina is a colony, usually of 8 or 16 cells, arranged in a solid sphere and surrounded by a **gelatinous envelope** (Fig. 7.2). Like *Gonium*, it is a free-swimming alga, each cell having two **flagella**.

Eudorina is also a spherical, free-swimming colony with a **gelatinous envelope.** However, it is hollow and has 16 to 64 cells (Fig. 7.3). Note the several **daughter colonies** present. The individual cells of both *Pandorina* and *Eudorina* are similar to those of *Chlamydomonas*.

Where appropriate, label the **gelatinous envelope**, a **daughter colony**, and **flagella** in the photographs of *Pandorina* and *Eudorina* (Figs. 7.2 and 7.3).

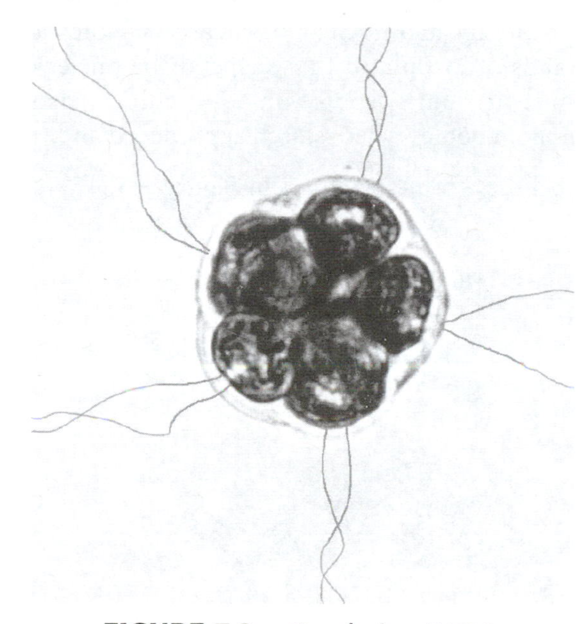

FIGURE 7.2 *Pandorina,* X 720.

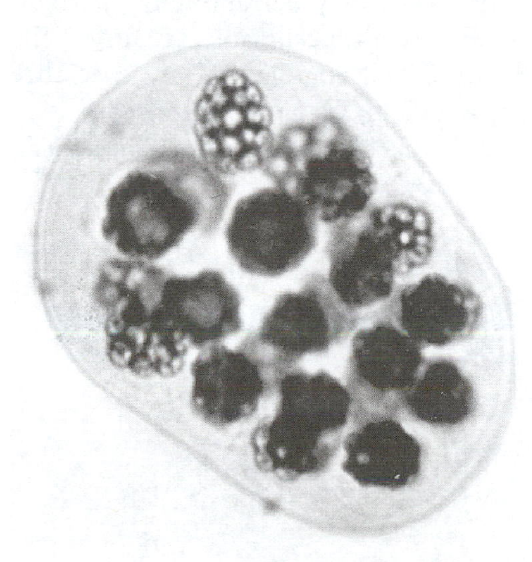

FIGURE 7.3 *Eudorina,* X 1050.

Volvox is the most complex member of this evolutionary line. The cells are connected by small cytoplasmic bridges (Fig. 7.4). Try to locate these connections when making your observations.

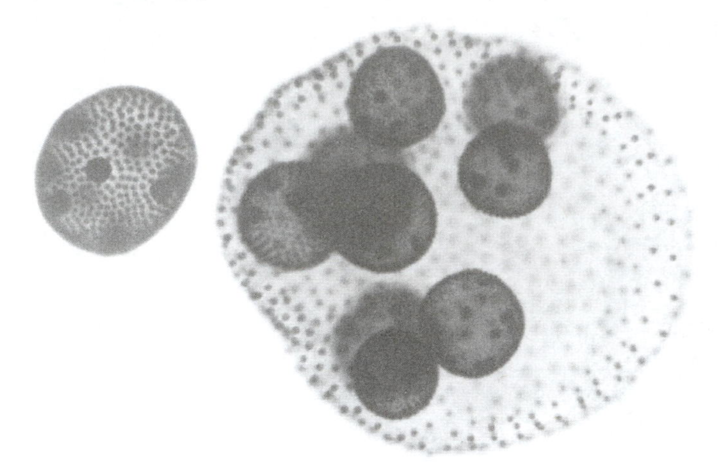

FIGURE 7.4 *Volvox*, showing early and later stages of daughter colony development, X 160.

Volvox colonies are so large that a special depression slide will be used for the observations. Place a small drop of the culture liquid in the depression. When a cover slip is added, a large air bubble should form in the center, surrounded by a ring of liquid containing the *Volvox* colonies.

Notice the spherical shape of the colony. The number of cells in a colony is variable, but there may be as many as 40,000 cells present.

Observe a single colony as it swims. Notice that the colony demonstrates a definite polarity (anterior and posterior) and that the cells in the posterior portion are slightly larger than those in the anterior part.

Some larger cells may be randomly distributed on the surface. The smaller cells are vegetative cells. The larger cells are called **gonidia** and are specialized for asexual reproduction. Examine several colonies to observe different stages of daughter colony formation. Several **daughter colonies** may develop within a single parent colony. Eventually, the older colony ruptures, releasing the daughter colonies.

1. Clean and return the depression slide.
2. *Volvox* also reproduces sexually. Examine a prepared slide to see this part of the life cycle. Only certain specialized cells are involved in this process. Typically, the sperm are produced in packets of 64 cells. This entire packet will swim to a mature female colony. A single sperm then fertilizes a mature egg. After fertilization, the zygote is retained within the female colony, and soon develops a thick, resistant cell wall. This thick-walled cell is referred to as a **zygospore**. This then goes through a period of dormancy.

Order Zygnematales: *Spirogyra*

Spirogyra is a very common alga of ponds and roadside ditches where there is only slight movement of the water. On warm, sunny days, photosynthesis often occurs so rapidly that oxygen bubbles get trapped by the algal net, raising the alga to the surface. The filaments are characteristically slippery or slimy to the touch (a fairly accurate way of identifying *Spirogyra* in the field).

1. Make a wet mount of *Spirogyra*. Notice the unusual, coiled shape of the **chloroplasts** from which the name of the alga is derived. Try to see the gelatinous covering also.
2. After making your initial observations, obtain a prepared slide of *Spirogyra*. This will show the single, centrally located **nucleus** of each cell. Note the fine strands of **cytoplasm** that suspend the nucleus. The numerous dark spots on the chloroplasts are **pyrenoid bodies** where starch is produced and stored.
3. Label these four structures in Figure 7.5.

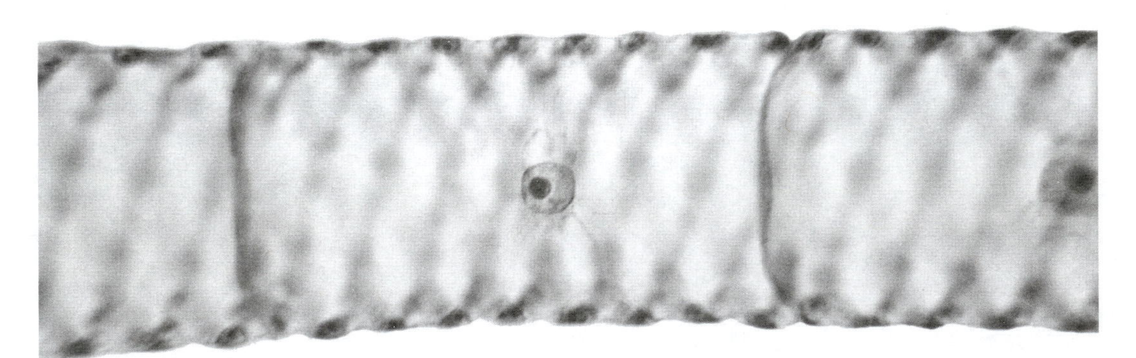

FIGURE 7.5 *Spirogyra*, vegetative cell, X 250.

Sexual reproduction in *Spirogyra* does not involve the formation of distinctive eggs and sperm. Instead, the entire contents of one cell passes through a special connecting tube and unites with the contents of another cell. This process is referred to as **conjugation**. After the nuclei join, producing a diploid zygote, the resulting cell usually develops a thick cell wall and enters into a dormant period as a **zygospore**.

Examine a prepared slide of *Spirogyra* in conjugation, and find the following stages of the process:

1. Cells of two adjacent filaments with the protuberances united forming a conjugation tube
2. Various stages of cytoplasmic transfer
3. Diploid zygotes and zygospores

When the zygospore germinates, the nucleus divides by meiosis producing four haploid cells. Three of these deteriorate; the remaining one then divides by mitosis to produce the vegetative filament. Label the photographs of *Spirogyra* in conjugation in Figure 7.6, indicating the **conjugation tube, motile gamete, nonmotile gamete, diploid zygote,** and **zygospore.**

Order Oedogoniales: *Oedogonium*

Obtain a prepared slide of *Oedogonium*, another example of a filamentous alga. Notice that the chloroplast has the appearance of a fine net within the cells.

Oedogonium can also be recognized by the presence of swollen **oögonia** (sing. = **oögonium**), which occur at intervals along the filaments. Each oögonium contains a single egg. Try to locate the **fertilization pore** through which the sperm enters.

The sperm are produced in smaller cells called **antheridia** (sing. = **antheridium**). These antheridia are produced in two ways, depending on the species. In the species we will study, they are produced in small clusters of 2 to 40 cells on the filament. The antheridia are very short, disklike cells, each producing two flagellated sperm cells.

Figure 7.7 is a photograph of *Oedogonium*. Label a **vegetative cell, antheridia, oögonium, oögonium with egg,** and **fertilization pore.**

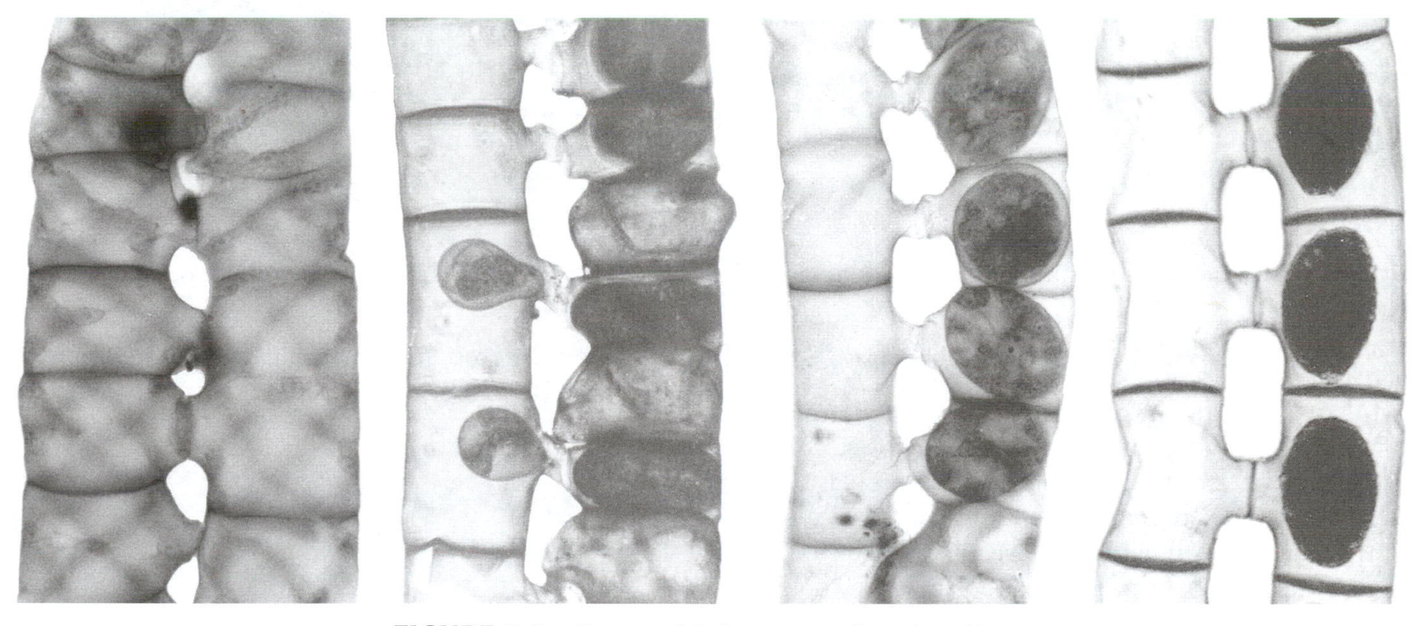

FIGURE 7.6 Stages of *Spirogyra* conjugation, X 100.

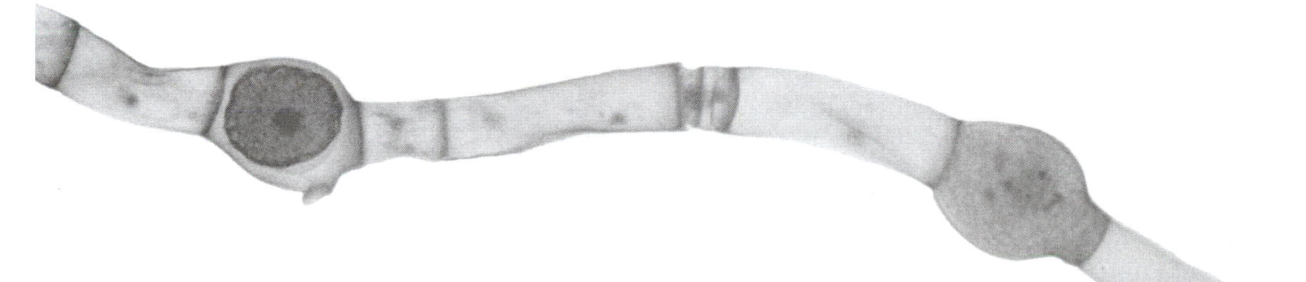

FIGURE 7.7 *Oedogonium,* X 360.

DIVISION CYANOBACTERIA (BLUE-GREEN ALGAE)

The blue-green algae are the most primitive members of the various algal groups. These are **prokaryotic** organisms, which means that their cells lack organized nuclei with nuclear membranes. Chloroplasts, mitochondria, and other membrane-bound cellular organelles are lacking as well. This primitive structure has resulted in modern classification of these organisms as bacteria. Until fairly recently, these were classified as **Cyanophyta**, a strange group of algae. For simplicity, I will continue to use the term *blue-green algae* for these organisms.

Photosynthetic pigments are present, however. The blue-green algae contain chlorophyll *a* (like all the other green plants) and accessory pigments called biliproteins. C-phycocyanin is a blue biliprotein, while *c*-phycoerythrin is a red biliprotein. Most of the blue-green algae have a predominance of *c*-phycocyanin, giving them their characteristic color. The main photosynthetic product is "cyanophycean starch," a carbohydrate that differs from the starch produced by the green algae and higher plants.

Flagellated cells have not been found in any species of blue-green algae. Any movement that occurs is by a characteristic gliding mechanism that is not completely understood as yet. Sexual reproduction has not been observed in these organisms either.

Gloeocapsa

Gloeocapsa is one of the simplest of the blue-green algae. It exists as either a unicellular form or as an aggregation of several cells (not a true colony, merely a cluster of cells).

Make a wet mount of the *Gloeocapsa* culture. Examine the cells with both low and high power. Notice the lack of internal organization (no chloroplasts, no nucleus, etc.). Find a cluster of cells and notice that each cell has its own gelatinous sheath, as does the entire cluster. With proper lighting, you should be able to see a series of **concentric gelatinous sheaths**.

Nostoc

Nostoc forms complex aggregations of many filaments. The generally spherical filaments may reach diameters of three inches or more. *Nostoc* may be found in quiet pools or on constantly moist surfaces, such as rock outcroppings or near the edge of waterfalls.

1. Observe the entire "colony." These "colonies" are composed of many filaments and a large amount of gelatinous material.
2. Squash a small portion of a colony on a slide and make a wet mount to observe the filaments. Notice that the filaments resemble a string of beads. Some spherical cells that are larger than the others may be seen. These are **heterocysts.** The formation of these cells weakens the filament, which may then break easily at that point. This is the major method of reproduction in *Nostoc* and is why many of the heterocysts are found at the end of filaments. In the blue-green algae, sections of a filament that result from fragmentation are known as **hormogonia.**
3. Label a **vegetative cell** and **concentric gelatinous sheath** in the following photograph of *Gloeocapsa* (Fig. 7.8).

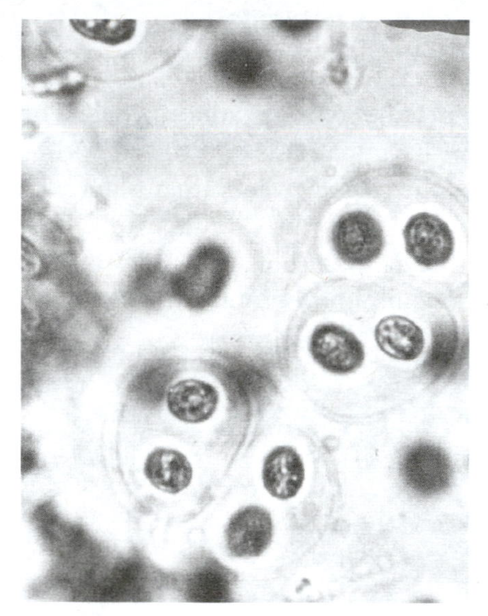

FIGURE 7.8 *Gloeocapsa,* X 650.

4. Label a **heterocyst** in the following photograph of *Nostoc* (Fig. 7.9).

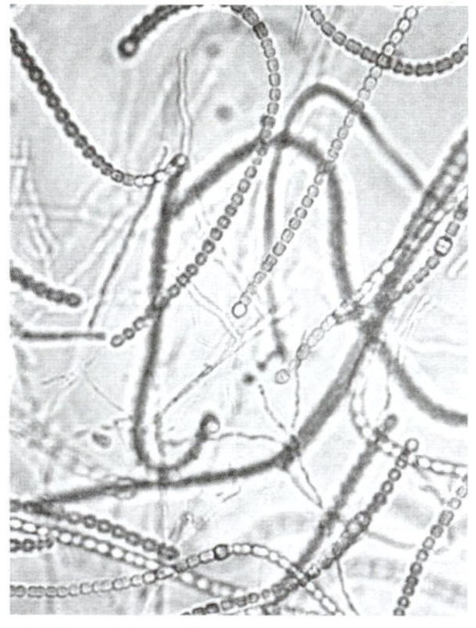

FIGURE 7.9 *Nostoc,* X 300.

Oscillatoria

Oscillatoria is a filamentous blue-green alga. It is very common in standing waters and is frequently found on the damp soil or pots in greenhouses.

1. Make a wet mount of the *Oscillatoria* culture. First examine it with low power, and note the swaying or oscillating motion of the filaments. Then examine a filament with high power. Notice that the individual cells are very short and disk shaped.
2. As in *Nostoc*, asexual reproduction in *Oscillatoria* involves the formation of hormogonia. These sections of the filaments develop by the formation of special cells called **separation disks**. These are essentially dead, clear cells along the filament. If these are not apparent, observe the prepared slide of a larger species. The separation disks will appear as biconcave cells along the filament.
3. Make a labeled diagram of *Oscillatoria* (Fig. 7.10), labeling the **vegetative cells** and **separation disks**.

FIGURE 7.10 *Oscillatoria.*

DIVISION EUGLENOPHYTA (EUGLENOIDS)

Members of the Euglenophyta are unicellular, flagellated organisms, some of which lack chlorophyll. The important diagnostic features of the division are (1) the presence of chlorophyll *a* and *b*, (2) the food reserve is known as paramylum, a polysaccharide that is different from starch, and (3) the cell wall is absent in this group. Instead of a rigid cell wall, the cell is bounded by a membranelike structure called the **pellicle**. The pellicle is composed of strips of proteinaceous material that surround the cell in a spiral manner. The pellicle may be very rigid or quite flexible.

One interesting feature of the euglenoid cell is the presence of an indentation at the anterior end. Nonpigmented forms are known to ingest microorganisms — a decidedly animal-like feature.

Euglena

1. *Euglena* is the best known member of this division. Make a wet mount of the culture, and find some cells that are not swimming about too actively.
2. Try to find the single, anterior **flagellum** that emerges from the anterior indentation or **reservoir**. How does the number of chloroplasts in *Euglena* compare with that found in *Chlamydomonas*? (1)

You should also be able to see shiny granules of **paramylum**, the stored food, and the red, light sensitive, **stigma**.

3. Make a diagram of *Euglena* in the space provided, labeling the **reservoir**, **stigma**, **flagellum**, **chloroplasts**, and **paramylum granules** (Fig. 7.11).

FIGURE 7.11 *Euglena.*

DIVISION XANTHOPHYTA (YELLOW-GREEN ALGAE)

At one time, the members of this division were included in the Chlorophyta. With more detailed studies, important differences between the two divisions became apparent: (1) there is no chlorophyll *b* present, and some species have chlorophyll *e* instead; (2) starch is never formed as a food reserve, but oils and fats are produced instead; (3) the cell wall is often absent, and if it is present, it often contains large amounts of silica; (4) the flagellated cells produced have two flagella of different lengths and types.

Vaucheria

Vaucheria is probably the best known of the yellow-green algae. It is commonly found in quiet ponds and may be found in moist terrestrial situations, such as those found in greenhouses.

1. Make a wet mount of the culture material, and note the tubular appearance of the filaments. The filaments of *Vaucheria* are essentially elongated multinucleate cells since no transverse cell walls are produced as the filament elongates. This tubular growth form is an example of **siphonous growth**, more technically known as **coenocytic growth.**

   Vaucheria reproduces sexually by oögamy. Each oögonium contains a single egg, and each antheridium, formed at the end of a curled filament, produces many sperm cells. The antheridia and oögonia are separated from the rest of the filament by cell walls. These are the only places where transverse cell walls are formed.
2. Examine the prepared slides of *Vaucheria geminata* and *V. sessilis* to see the sexual structures of *Vaucheria*. These may be present in the culture but generally are not.
3. Label an **oögonium** and an **antheridium** in *each* of the following photographs of *Vaucheria* (Figs. 7.12 and 7.13).

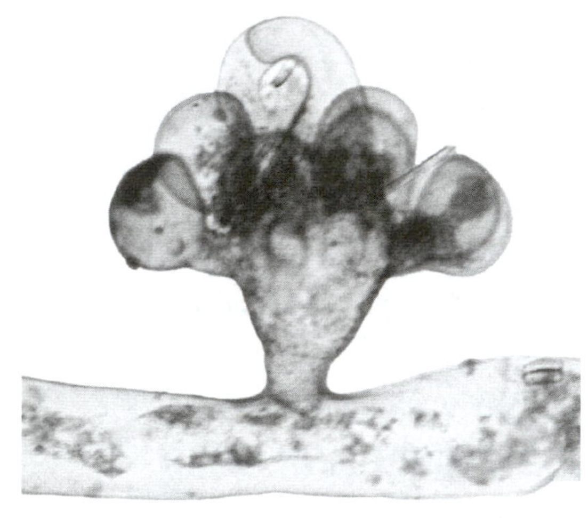

FIGURE 7.12 *Vaucheria geminata,* X 155.

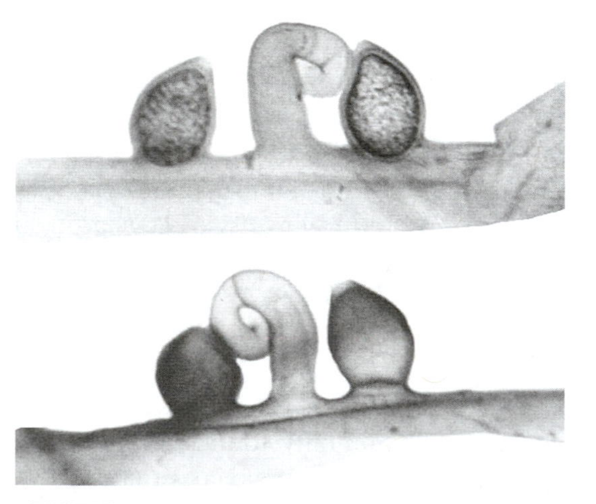

FIGURE 7.13 *Vaucheria sessilis,* X 145.

DIVISION BACILLARIOPHYTA (DIATOMS)

Diatoms comprise the most important group of phytoplankton. These are small, single-celled organisms (in some species the cells remain attached to other cells after cell division, resulting in short filaments or other arrangements). The cells are heavily impregnated with silica and often have very fine, delicate markings. (Cleared diatom shells were once used as a subjective evaluation of the quality of microscope objectives.)

Chlorophyll *a* and *b* are present, but their green colors are masked by the presence of a brown pigment called fucoxanthin. Starch is not produced, but oils are the important food reserves.

While some freshwater diatoms are planktonic, most are found attached to submerged objects or live on the mud of the ponds.

1. If available, examine a drop of the culture of diatoms. Take your sample from the bottom of the container. The diatoms will be small, generally rectangular cells. Prepared slides of mixed diatoms are available for observation.
2. Diagram three different species of diatoms in the following space (Fig. 7.14), showing the markings on the silica cell wall. Try to identify the genus of each by using the available reference books.

FIGURE 7.14 Miscellaneous diatoms.

3. Observe the demonstration of specially prepared diatom cells. Here the cytoplasm has been removed and only two or three cells have been placed on the slide and arranged manually. Please be careful.

Questions

1. Why should the term *algae* be one of convenience rather than one with definite technical connotations? (2)

2. Why could conjugation in *Spirogyra* be considered a form of isogamy? (3)

3. What is oögamy? (4)

4. How could you tentatively identify *Spirogyra* if you had no microscope? (5)

5. What is one possible function of a zygospore? (6)

   a. When would algae most likely produce these? (7)

6. Why are the Cyanobacteria considered the most primitive in the algae? (8)

7. What is the major (if not the only) method of reproduction in the Cyanobacteria? (9)

8. Using features that you observed in lab, explain how *Chlamydomonas* differs from *Euglena* in appearance? (10)

Chlamydomonas	*Euglena*

9. Does the species of *Euglena* seen in class have a rigid or flexible pellicle? (11)

10. How can diatoms be recognized easily? (12)

Fungi: Part 1

Introductory Notes

Fungi, like the algae, represent a heterogeneous assemblage of thallus plants — plants that lack roots, stems, and leaves. Unlike the algae, fungi cannot carry on the process of photosynthesis. All fungi are heterotrophs. Because of this difference and because of some unusual cell structures, the fungi are now placed in their own kingdom. We will consider them as unusual plants, comprising four divisions. (Some texts list only two divisions and others as many as six.)

Just as the algae are extremely important as primary producers in many food chains, the fungi (in concert with bacteria) are equally important at the other end of food chains, i.e., acting as decomposers. As such, they bring about the final conversion of complex organic compounds to simpler substances that can be reused by green plants, entering the food chain again. The fungus itself may be eaten and, in this way, start a food chain. Quite often, this decomposition activity has a direct effect on humans through wood rotting, food spoiling, leather molding, or some other detrimental action.

In addition to their importance as decomposers, many fungi are parasites on both animals and plants. Some are dangerous pathogens. Other fungi are extremely beneficial to man as sources of food, antibiotics, enzymes, etc. The baking and brewing industries could not survive without the activities of certain fungi. The flavorings in various cheeses are the direct result of fungal growth.

GOALS

After completing this exercise, the student should be able to:

- explain how fungi differ from other plants.
- differentiate between slime molds and true fungi.
- list at least two characteristics of the division Phycomycota.
- identify the genera *Allomyces*, *Saprolegnia*, and *Rhizopus* from either a photograph or a diagram.
- define or identify the various structures of the genera used in this exercise. These include plasmodium, myxomycete, fungus, ectoplasm, endoplasm, capillitium, seta, columella, hypha, mycelium, rhizoid, gametophyte, gametangium, sporophyte, sporangium, zoöspore, coenocytic, septate, sporangiophore, stolon, suspensor, zygospore, homothallic, heterothallic, conidia, and conidiophore.
- identify the genera *Aspergillus* and *Penicillium* from either a photograph or a diagram.
- answer the questions in the exercise.

DIVISION MYXOMYCOTA (SLIME MOLDS)

Slime molds are very inconspicuous fungi. The vegetative part of a slime mold is called a **plasmodium.** The plasmodium is a slimy, naked mass of protoplasm with many diploid nuclei. Quite often these can be found under the bark of fallen, partially rotten logs. The plasmodium has amoeboid movement, and as it flows or creeps around, it engulfs microorganisms. This phase of the life cycle is distinctly animal-like, and in fact, slime molds are sometimes studied in protozoology classes.

Examine the demonstration petri plates containing the vegetative stage of the slime mold *Physarum.* Notice the streaming of the protoplasm in the plasmodium strands. The outer, more dense portion of the plasmodium is called the **ectoplasm,** while the inner, more fluid part is the **endoplasm.** Where is the protoplasmic steaming most evident, in the central portion or at the periphery? (1)

Some spore-producing stages of a few different species are on demonstration. Examine these and note their very inconspicuous nature. Please do not handle them, they are very fragile. (In the slime molds and in most other fungi, the spore-producing stage is known as a *fruiting body*.)

Obtain a prepared slide of the spore-producing stage of the slime mold *Stemonitis*. Examine this with a dissecting microscope. The basal stalk is called the **seta.** The basal portion of the seta is the **holdfast.** The fine network of threads that retains the spores is called the **capillitium.** The extension of the seta through the capillitium is the **columella.**

Label these on the photograph of *Stemonitis* (Fig. 8.1).

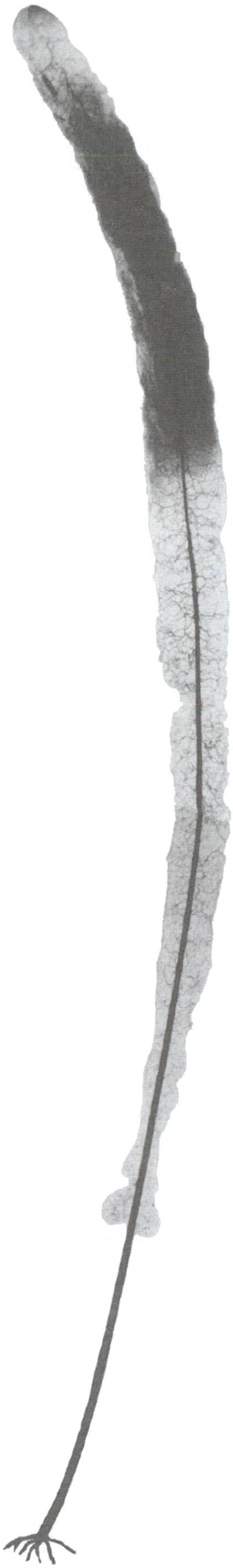

The remaining fungi are sometimes placed in a single division, the Eumycophyta or "true fungi" by some mycologists (specialists in the study of fungi). The main reason for this grouping is that the plant body of these fungi is composed of branching, filamentous filaments called **hyphae** (sing. = **hypha**). These frequently occur in an intertwining network of various degrees of complexity and density. This mass is known as a **mycelium**. The hyphae contain nuclei and may be either septate or nonseptate. Other mycologists believe that the various groups are sufficiently dissimilar to warrant placing them in separate divisions. This is the approach that will be followed here.

FIGURE 8.1
Fruiting body of *Stemonitis*, X 25.

DIVISION PHYCOMYCOTA (ALGAL FUNGI)

These organisms are called *algal fungi* because they have many similarities to filamentous green algae: many are aquatic; the mycelium is a loose network of hyphae; and the methods of reproduction are similar. They, of course, do not photosynthesize, and any evolutionary relationship is most certainly open to question. In fact, Phycomycota is rarely used as a taxonomic unit now. These organisms are more frequently classified as three separate divisions, two of which are in the kingdom Protista.

The hyphae of these fungi lack transverse cell walls and are multinucleate (coenocytic). Some members of the group are major parasites, but most are probably important as decomposers. Three members of the group will be examined to show some of the diversity and different life histories present in the division.

The three genera that will be studied here are treated as members of three classes. *Allomyces* is a representative of the class Chytridiomycetes, *Saprolegnia* represents the Oömycetes, and *Rhizopus* the Zygomycetes. Many authorites have now elevated these classes to division status, and no longer recognize Phycomycota as a valid taxonomic unit.

Allomyces

Allomyces is an aquatic fungus that differs from most other fungi in that it has a distinct alternation of generations. Both stages produce extensive, dichotomously branched hyphae.

The **gametophyte**, or sexually reproducing, stage produces both male and female **gametangia** (sing. = **gametangium**) on the same plant. Usually, these are produced in pairs at the tips of the hyphal branches.

The male gametangia typically become orange due to the production of a carotenoid pigment during the development of the sperm.

Both types of gametes are released into the water, where fertilization occurs. In *Allomyces*, the female gamete is larger than the male gamete. This situation is referred to as **anisogamy**.

Make a wet mount of some of the gametophyte material, and examine it with both low and high power. Locate a region where you can see both types of gametangia. Which type is located at the very tip of the hypha, the male or the female? (2)

Make a labeled diagram of the gametophtye stage of *Allomyces* (Fig. 8.2). Show and label the **male** and **female gametangia** and their attachment to a **hypha**.

FIGURE 8.2 *Allomyces* gametophyte.

The **sporophyte** stage develops from the zygote and it is diploid in nature. When it is mature, it produces oval **zoösporangia** (sing. = **zoösporangium**) at the tips of the hyphae and at the ends of short, lateral branches. Two types of zoösporangia are produced. Those that are produced early in the development of the fungus are thin-walled and colorless, while those produced later have thicker, darker walls.

The thin-walled zoösporangia produce flagellated **zoöspores** that will give rise to new sporophyte plants. What type of cell division is involved in the formation of the zoöspores in these thin-walled zoösporangia, meiosis or mitosis? (3)

Make a wet mount of some of the sporophyte material, and examine it as you did the gametophyte. Which type of zoösporangium is most abundant? (4)

Make a labeled diagram of the sporophyte stage of *Allomyces* (Fig. 8.3), showing the shape and location of the **zoösporangia** on the **hypha**.

FIGURE 8.3 *Allomyces* sporophyte.

Saprolegnia

Saprolegnia is a common water mold. It is usually a saprophyte but has been known to parasitize fish or fish eggs.

Make a wet mount of the *Saprolegnia* mycelium, and examine it first with low power. Both sexual stages and asexual stages can be found on the same mycelium. Locate some of the elongated, cylindrical zoösporangia. These will produce flagellated zoöspores that will be released into the water and develop into new plants. Is this an example of sexual or asexual reproduction? (5)

Saprolegnia is **homothallic**; both male and female gametangia (here referred to as antheridia and oögonia) develop on the same mycelium and are self-compatible. Only one strain is needed for sexual reproduction. The gametangia are produced at the ends of short, lateral branches. The oögonia are spherical cells and contain several eggs. The antheridia are slender and multinucleate and are usually produced near an oögonium. The antheridia have several branches that eventually come in contact with an oögonium. Slender hyphae called *fertilization tubes* then penetrate the oögonium, and each comes in contact with an egg. Male gametes then migrate through the tubes and fertilize the eggs.

Why are the sexual reproductive structures referred to as antheridia and oögonia instead of gametangia? (6)

Make a labeled diagram of *Saprolegnia*, showing an **oögonium** with **eggs** and attached **antheridium** (Fig. 8.4).

FIGURE 8.4 Sexual reproductive structures of *Saprolegnia*.

Rhizopus

Rhizopus stolonifer is a very common fungus, commonly called *black bread mold*. Its spores are quite abundant in the atmosphere, and they develop on a variety of substrates (homemade bread or rolls with no preservatives is always a good place). Like other phycomycetes, the hyphae are non-septate and multi-nucleate.

Use a dissecting microscope to examine a petri plate containing a culture of *Rhizopus*. Note the white mycelium and the numerous black spherical structures. These black structures are the sporangia. Careful observation should reveal the elongated stalk of each sporangium and that these stalks are connected to others by a slender horizontal hypha.

Make a wet mount of some of the culture material. It is not necessary to get a large amount of the culture. If necessary, use your text to locate the following structures:

1. The **sporangiophore** is the aerial hypha that produces and elevates the **sporangium** (pl. = **sporangia**).
2. Examine the sporangium for **spores** and the central **columella.** When making a wet mount, the thin wall of the sporangium usually ruptures, and all that can be seen is the columella with some spores attached.
3. **Stolons** are elongated, horizontal hyphae that grow over the surface of the medium, connecting one cluster of sporangiophores to another.
4. **Rhizoids** are short, branching hyphae at the base of the sporangiophore. What are two possible functions of these structures? (7)
5. Label the **sporangiophore, sporangium with spores, columella, stolon,** and **rhizoids** in the photograph of *Rhizopus* (Fig. 8.5).

Rhizopus is a **heterothallic** fungus (two different strains are required for sexual reproduction). Since the mycelia of the two strains are identical in appearance but are physiologically different, they are referred to as + and − strains. When two compatible strains are in close proximity to each other, protuberances referred to as **progametangia** develop from each hypha. Soon the tips of these protuberances become separated from the rest of the hypha by a cell wall. This results in a cell called a **gametangium.** The cell wall between the two gametangia break down, and there is a mixing of the protoplasts. The nuclei then fuse, resulting in a diploid **zygote,** which soon develops a thick cell wall and is then known as a **zygospore.** The portion of the hypha connected to the zygospore is the **suspensor.**

To complete the life cycle, the diploid cell divides by meiosis. The resulting haploid nuclei produce the hyphal filaments such as was seen in the first part of this exercise.

Examine the culture of + and − strains. Find the line of zygospores that form at the point where both strains came into contact. Make a wet mount of this material.

Examine a prepared slide of *Rhizopus,* and try to find as many of the previously mentioned structures as possible. Label a **progametangium, gametangium, zygote, zygospore** and **suspensor** in Figure 8.6.

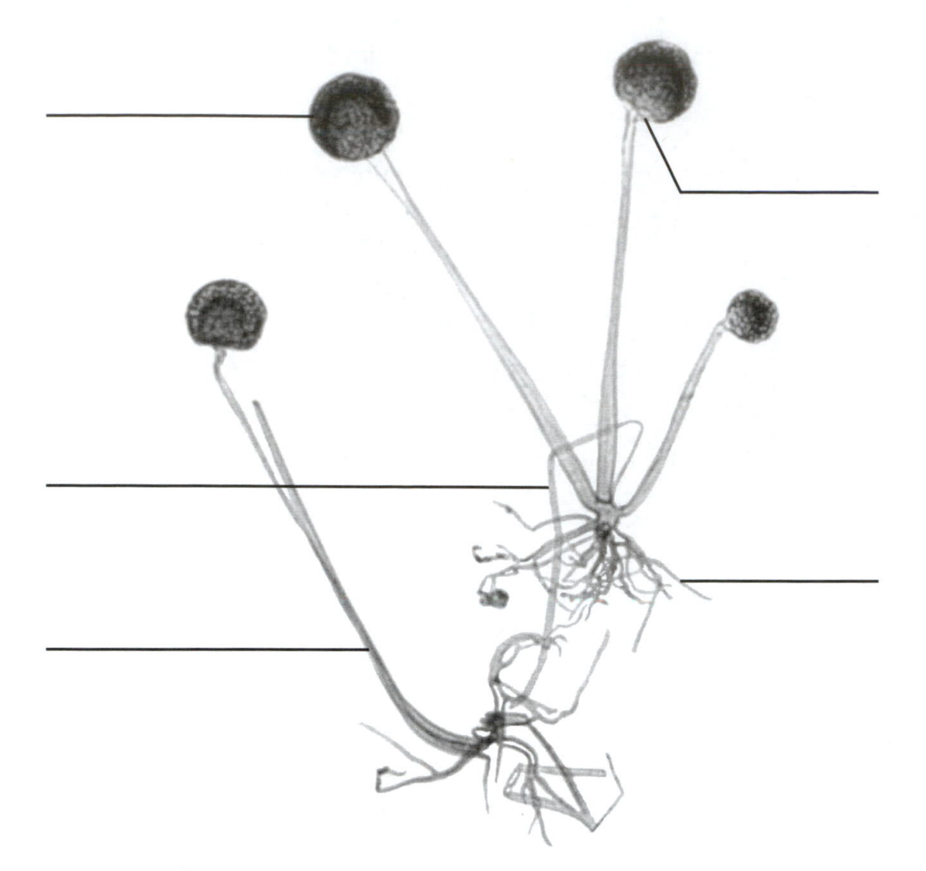

FIGURE 8.5 *Rhizopus stolonifer*, asexual reproduction, X 300.

Courtesy Carolina Biological Supply Company

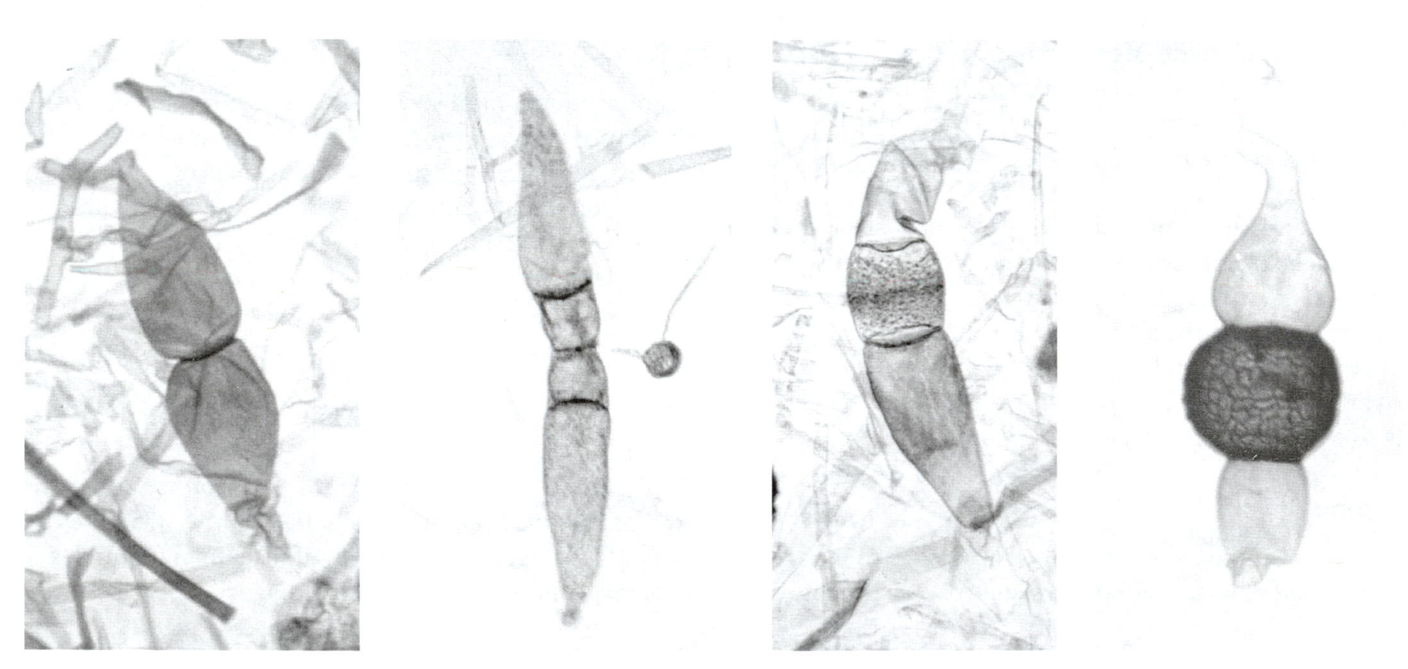

FIGURE 8.6 Stages of sexual reproduction in *Rhizopus;* X 180, X 210, X 185, X 150.

DIVISION DEUTEROMYCOTA (FUNGI IMPERFECTI)

This is a large group of fungi that have one unusual thing in common: the sexual (or perfect) stage is unknown; all reproduction is asexual. It may be that some have not been studied very extensively and the sexual stages not found. It also appears that the sexual stage has been lost by many. We will examine two common genera: *Aspergillus* and *Penicillium*.

Common Brown and Green Molds

These fungi are extremely common and widespread. They grow on many substrates, such as cheese, oranges, bread, leather, etc. The typical color of the fungus is caused by the presence of numerous asexual spores called **conidia** (sing. = **conidium**). These conidia are produced at the ends of aerial hyphae called **conidiophores**. The conidia are produced in chains by a pinching in or constricting of the end of the conidiophore.

Make wet mounts of these molds and locate the conidiophores. Notice that the conidiophores of *Aspergillus* are produced on a spherical swelling at the end of an elevated hypha, while the conidia arrangement of *Penicillium* resembles a small whisk broom.

Label the **conidia** and **conidiophores** in Figures 8.7 *and* 8.8.

FIGURE 8.7 *Aspergillus* conidiophore, X 580.

FIGURE 8.8 *Penicillium* conidiophore, X 400.

Questions

1. Slime molds are sometimes studied in zoology courses. List two features of slime molds that could be considered animal-like and two plantlike features. (8)

Animal-like	Plant-like
1.	1.
2.	2.

2. How do the slime molds differ from the true fungi? (9)

   Slime molds:

   True fungi:

3. Which alga, studied recently, has a pattern of reproduction similar to that of *Saprolegnia*? (10)

4. In *Rhizopus*, the sporangia are produced at the end of sporangiophores. What is a possible benefit of this arrangement? (11)

5. What is the difference between a parasite and a saprophyte? (12)

   Parasite:

   Saprophyte:

6. *Saprolegnia* was said to be homothallic; *Rhizopus* was said to be heterothallic. What do these two terms mean? (13)

   Homothallic:

   Heterothallic:

7. How does spore production of *Rhizopus* differ from conidia production in *Aspergillus* or *Penicillium*? (14)

8. How are conidia and spores similar? (15)

Fungi: Part 2

Introductory Notes

DIVISION ASCOMYCOTA (SAC FUNGI)

Class Ascomycetes

Most ascomycetes have well-developed mycelia, the hyphae of which usually have cross walls. The most distinguishing feature of these organisms is the presence of a saclike structure called an **ascus** (pl. = **asci**) in which **ascospores** are produced. Ascus and ascospore production follow sexual reproduction. The diploid zygote divides by meiosis, producing four haploid cells. Typically, these each divide once by mitosis, resulting in the production of eight haploid ascospores. These are usually found inside a fruiting body called an **ascocarp**. Three different types of ascocarps are recognized:

1. **Apothecium** (pl. = **apothecia**), an open, cup-shaped fruiting body.
2. **Cleistothecium** (pl. = **cleistothecia**), a closed, spherical fruiting body.
3. **Perithecium** (pl. = **perithecia**), a flask-shaped fruiting body with a pore (the ostiole) at the tip.

In addition to the development of ascospores, many ascomycetes also produce asexual spores or conidia. Very likely, some of the Deuteromycota studied in the previous exercise are ascomycetes that have lost the ability to reproduce sexually.

GOALS

After completing this exercise, the student should be able to:

- identify the various genera studied from either a photograph or a diagram.
- explain the distinguishing features of both the Ascomycota and the Basidiomycota.
- describe and identify the three types of ascocarps.
- make a labeled diagram of a mushroom.
- make a labeled diagram of a portion of a mushroom gill, showing basidia, sterigmata, and basidiospores. Identify these from a photograph or a diagram.
- recognize and identify the following stages of wheat rust: aecia, aeciospores, uerdia, urediospores, telia and teliospores, and pycnia.
- make a diagram of the wheat rust life cycle.
- explain the significance of mycorrhizae, and differentiate between endomycorrhizae and ectomycorrhizae.
- identify the three main growth forms of lichens.
- answer the questions in the exercise.

Brewer's Yeast (*Saccharomyces cerevisiae*)

Yeasts are single-celled fungi that do not produce ascocarps. The most common type of reproduction is by an unusual form of cell division called **budding**. The nucleus divides by mitosis, and at the same time, a small protrusion develops on the cell wall of the parent cell. This protrusion enlarges, and the daughter nucleus migrates into this new **bud**. Examine a drop of *Saccharomyces cereviciae* and locate cells that are in the process of budding. Chains of cells may be produced if budding is rapid.

Make a diagram of budding yeast cells, labeling the **parent cell** and the **bud** (Fig. 9.1).

FIGURE 9.1 *Saccharomyces* budding.

Powdery Mildew of Lilac (*Microsphaera alni*)

The powdery mildews are a fairly large group of plant parasites. Some are extremely destructive, while others, such as *Microsphaera alni*, seem to do little harm to their host.

Examine a demonstration leaf that has been set up under a dissecting microscope. The white, powdery appearance is due to the extensive growth of the mycelium and the production of many conidia. The small, dark dots are the ascocarps.

Using another leaf, scrape some of the ascocarps into a drop of water on a slide, and make a wet mount. Use the compound microscope and examine these with low power. What type of ascocarp is this? (1)

Notice the external appendages attached to the ascocarp. These are characteristic for a genus and are useful in identification. Those of *Microsphaera* are the most complex. What might be one possible function of these appendages? (2)

While you are examining the ascocarps through the microscope, press gently on the cover slip with a dissecting needle until the ascocarp ruptures. This will force the asci out. Notice that more than one ascus is present. Usually, eight ascospores are present, but these have no particular pattern of organization.

Figure 9.2 is a photograph of a ruptured fruiting body of *Microsphaera alni.* Label the **ascocarp** (by type), **appendages, ascus,** and **ascospores.**

Sordaria

Sordaria fimicola is a common and extensively studied species of ascomycete. A saprophyte, it normally grows on decaying plant material or dung. It does not produce any asexual spores, reproducing solely by ascospore formation.

Make a wet mount of some of the mycelium of *Sordaria* from the culture dish. Be sure to get some of the dark brown fruiting bodies. Examine the ascocarps with low power. What type of ascocarp does *Sordaria* have? (3)

While you are examining an ascocarp, press gently on the cover slip with a dissecting needle to force some of the asci out. Notice that the asci are elongated, and each contains eight ascospores arranged in a linear pattern. Careful examination of an ascus will reveal a small pore at the tip through which the ascospores are released. This linear arrangement of the ascospores has proved beneficial for genetics studies; individual ascospores can be observed, removed, and cultured.

Label the **ascocarps** (by type), **ascus,** and **ascospores** in the following photographs of *Sordaria* (Figs. 9.3 and 9.4).

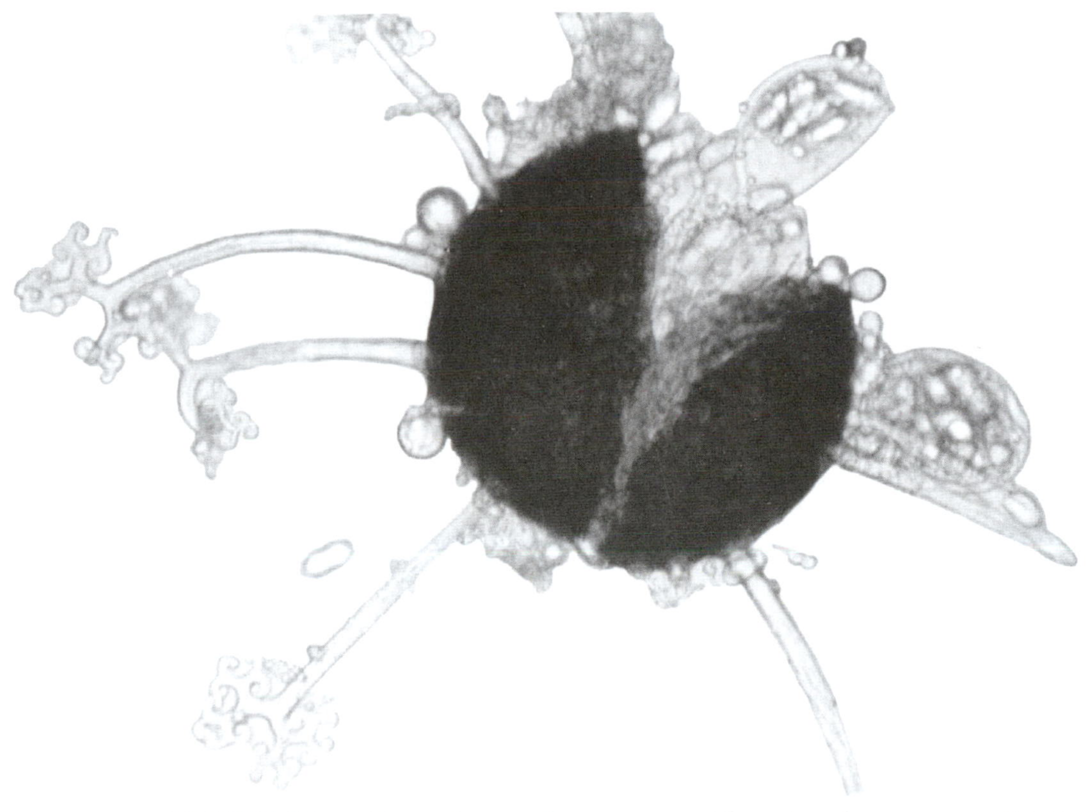

FIGURE 9.2 *Microsphaera alni* ascocarp with asci and ascospores, X 400.

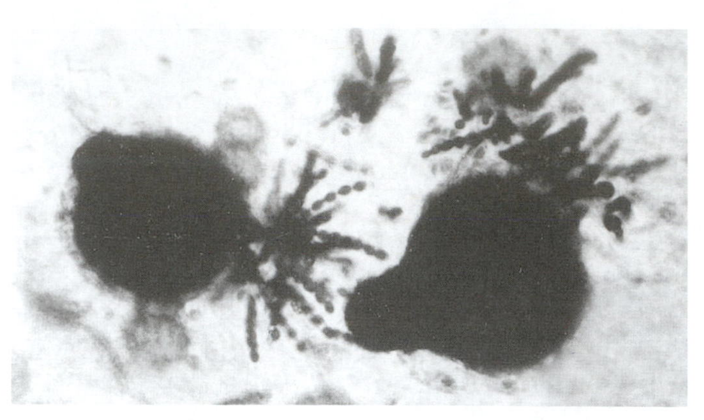

FIGURE 9.3 *Sordaria* ascocarp, X 380.

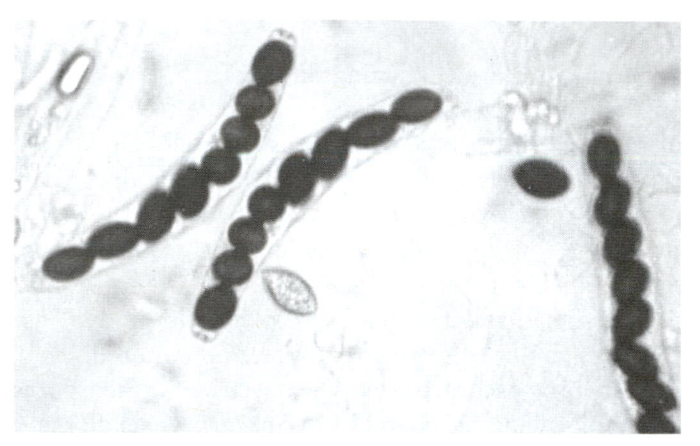

FIGURE 9.4 *Sordaria* asci and ascospores, X 1100.

Cup Fungus (*Peziza sp.*)

Examine the preserved specimens and then a prepared slide of a longitudinal section through the spore-producing structure of *Peziza*. Notice that the elongated asci form a very obvious layer. This layer is referred to as a **hymenium**. What type of ascocarp does *Peziza* have? (4)

Label the **ascocarp** (by type) and **hymenium with asci** in the photograph of *Peziza* (Fig. 9.5)

Ergot of Rye (*Claviceps purpurea*)

Claviceps is an interesting and dangerous parasite of cereal grains. The fungus invades the seed-producing heads of the plant and forms a hard, black mass known as an **ergot**. Several toxic substances are produced in this ergot (one of which is used to produce LSD). If these ergots are ground up in flour, the toxins may be passed on to people. The symptoms of ergot poisoning are numerous, including nervous spasms, hallucinations, and temporary insanity. One common effect is a constriction of the blood vessels, especially in the extremities. This causes the limbs to become swollen and inflamed; burning sensations alternate with those of

FIGURE 9.5 Longitudinal section through the fruiting body of *Peziza,* X 95.

extreme cold. The limbs may become numb, shrunken, and mummified. Occasionally, entire nails, toes, fingers or larger units may be shed. One severe epidemic in the Near East, in A.D. 944, is estimated to have killed some 40,000 people. There are still occasional, less severe, reports of ergotism.

This ergot also overwinters in or on the soil, or fields may be sown with seeds containing ergots. In the spring, these germinate and produce elongated structures with spherical tips called **stromata** (sing. = **stroma**). In the spherical tips of the stroma are numerous ascocarps with elongated asci, each with eight linear ascospores. These ascospores then infect the new wheat crop.

1. Examine the demonstration of wheat infected with *Claviceps* to see the ergot.
2. Examine a prepared slide of a longitudinal section of a stroma to see the ascocarps and linear ascospores. Individual asci are not visible on these slides.
3. What type of ascocarp does *Claviceps* have? (5)
4. Label the **stroma** and **ascocarp** (by type) of *Claviceps* (Fig. 9.6).

FIGURE 9.6 Longitudinal section through a stroma of *Claviceps purpurea,* X 55.

DIVISION BASIDIOMYCOTA (CLUB FUNGI)

Class Basidiomycetes

The basidiomycetes also have hyphae with transverse cell walls. However, these fungi are characterized by sexual spores that are produced externally on a club-shaped structure called a **basidium** (pl. = **basidia**). The **basidiospores** are attached to the basidia by short stalks called **sterigmata**. The basidia are usually produced on a larger fruiting body called the **basidiocarp**. Basidiocarps are probably the most familiar fungi, including mushrooms, puffballs, and bracket fungi. Nearly all of the fungi just listed are saprophytes. Other basidiomycetes, the rusts and smuts, are harmful plant parasites. These do not produce well-developed basidiocarps.

Cultivated Mushroom (*Agaricus sp.*)

The common cultivated mushroom, purchased in a grocery store, is only a small portion of the entire fungus body. (The same thing can be said for mushrooms or "toadstools" that grow on one's lawn.) All that is picked is the basidiocarp or fruiting body. The mycelium has been growing for some time in the soil (or other medium).

Examine some basidiocarps of different ages and find the following structures: the **pileus,** or cap; the **stipe,** or stem; the **inner veil** (a membrane that attaches the pileus to the stipe), and the **annulus**, or ring that forms around the stipe when the pileus breaks away from the stipe. The annulus will be found only on more mature mushrooms. When the pileus does break away from the stipe, the **lamellae**, or **gills**, are exposed. The basidia and basidiospores are produced on the surface of the gills.

1. Cut a mushroom in half, longitudinally, and make a labeled diagram showing *either* the **inner veil** *or* the **annulus** (Fig. 9.7).
2. Make a wet mount of a thin section of part of the pileus or stipe to examine with the compound microscope. Note the many filaments (hyphae) that comprise the mycelium of the basidiocarp.
3. Examine a portion of a gill and try to locate the basidiospores.

FIGURE 9.7 Longitudinal sections of *Agaricus.*

Shaggy-mane Mushroom, etc. (*Coprinus sp.*)

Obtain a prepared slide of *Coprinus*. This is a cross section of the pileus. Note the arrangement of the gills and how they resemble that of *Agaricus*. This radiating pattern is very common. Examine the slide first with low power, and find a region where the tiny basidiospores are evident. Switch to high power and note the structure of a **basidium.** It is somewhat club-shaped, sticking out from the surface of the gill. Look at the free ends of the basidia to find the **sterigmata** (sing. = **sterigma**) with **basidiospores** attached.

Label the **stipe, pileus,** and a **lamella** in the photograph of *Coprinus* (Fig. 9.8).

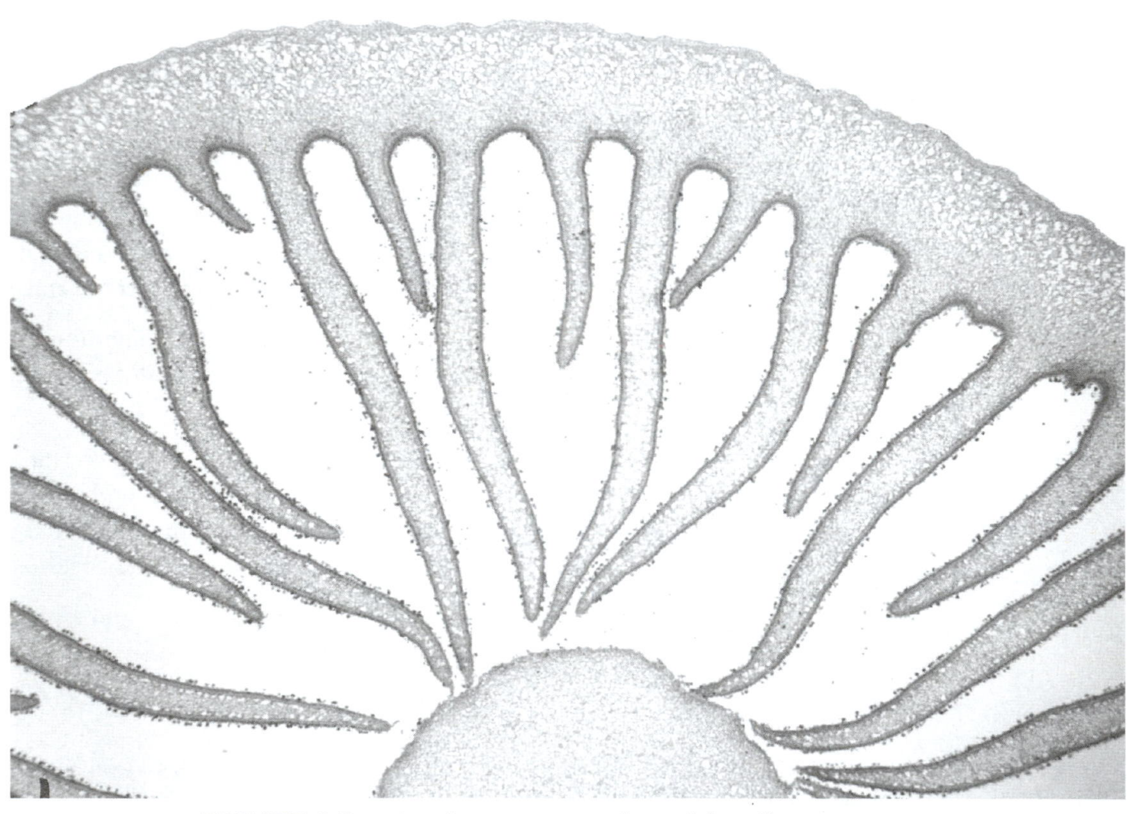

FIGURE 9.8 *Coprinus*, cross section of the pileus, X 35.

Label a **basidiospore, sterigma, basidium,** and the **lamella** in the close-up photograph of *Coprinus* (Fig. 9.9).

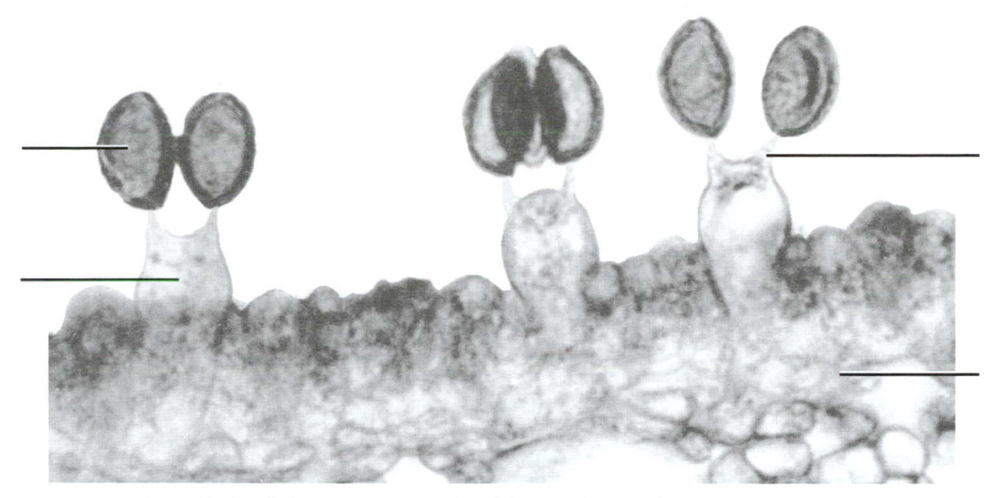

FIGURE 9.9 *Coprinus*, basidia with basidiospores, X 850.

Wheat Rust (*Puccinia graminis*)

There are several varieties of *P. graminis*, all of them serious pathogens of grains (wheat, barley, oats, rye, etc.). This is an extremely specialized parasite, requiring an alternate host for the completion of its life cycle and producing five kinds of spores.

Obtain a prepared slide of *P. graminis*. This slide has most of the stages of the life cycle on it. There are two sections of wheat and one section of the leaf of the alternate host, the barberry plant, which is a dicot. Be sure you can identify the following structures and know the sequence of the life cycle.

1. **Uredospores** are found on the wheat plant in blisterlike clusters called **uredia** (sing. = **uredium**). These uredia form rusty brown patches that give the fungus its name. Uredospores are blown by the wind and can infect other wheat plants. Since wheat plants are grown close together, this spreads the disease rapidly. The uredospores are single-celled and have relatively thin cell walls.

   Label the **uredospores**, the wheat **epidermis**, and a **vascular bundle** shown in Figure 9.10.

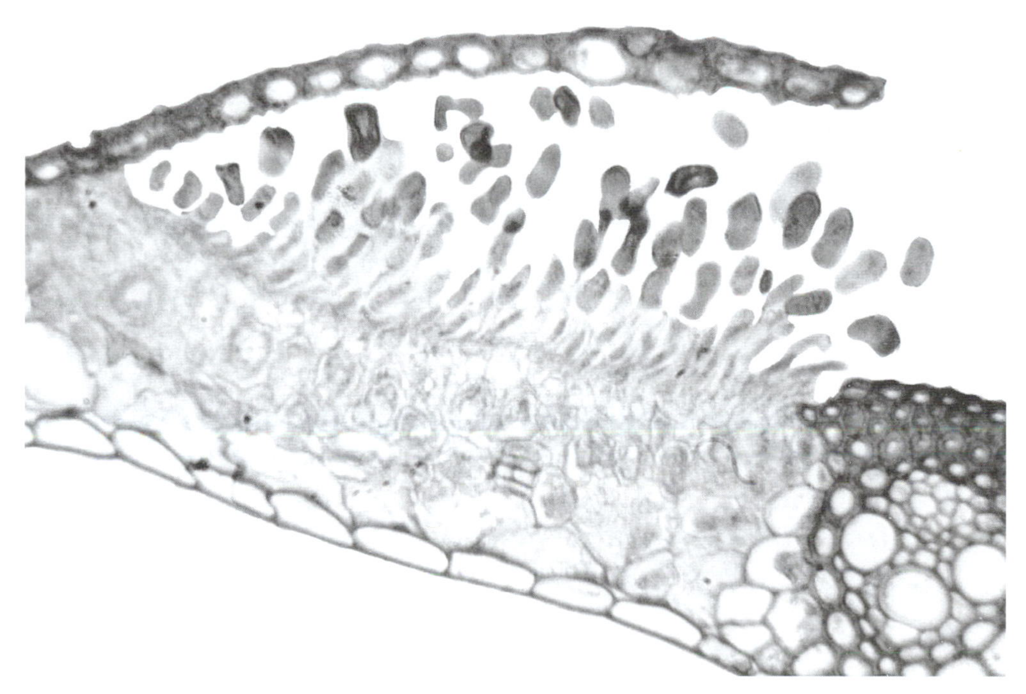

FIGURE 9.10 *Puccinia graminis* uredium on wheat, X 275.

2. Examine the other wheat stem cross section to find **teliospores**. These are produced in late summer when the wheat is mature. The two-celled teliospores are produced in clusters called **telia** (sing. = **telium**), which appear black on the stem of the plant. Unlike the uredospores, teliospores have a thick cell wall and are two-chambered structures.

   Label the **teliospores**, the wheat **epidermis**, and a **vascular bundle** shown in Figure 9.11.

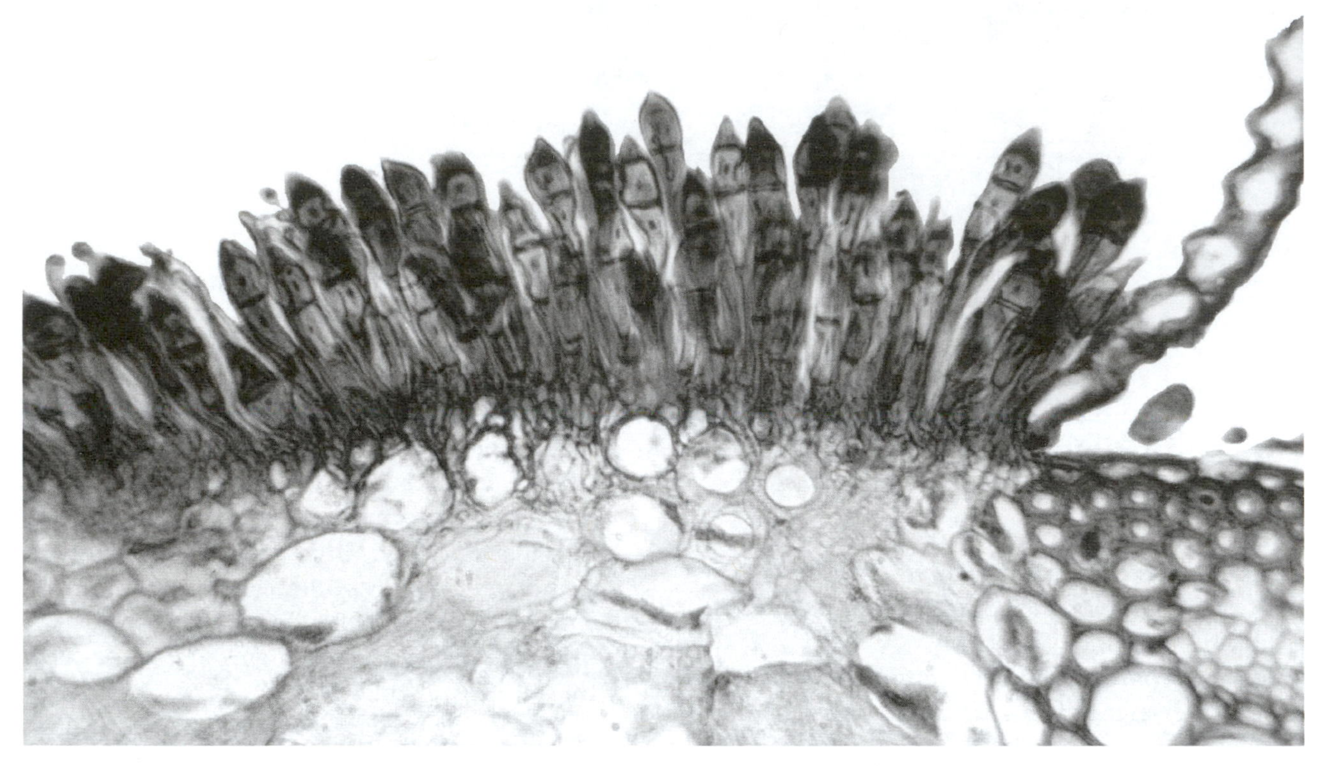

FIGURE 9.11 *Puccinia graminis* telium on wheat, X 310.

3. Teliospores do not reinfect other wheat plants, but overwinter on the ground or in dead straw in fields. In the spring, each of the two cells produces a basidium that produces four basidiospores. *These basidiospores can only infect the alternate host, the barberry plant.*

   We do not have the basidiospore stage of the life cycle.

4. In the leaf of the barberry plant, a mycelium develops. Soon, clusters of hyphae develop on the upper surface of the leaf. Each of these is called a **spermagonium** or **pycnium**. This is where sexual reproduction is initiated.

5. Soon after sexual reproduction, bell-shaped clusters of spores develop on the lower surface of the barberry leaf. These are called **aecia** (sing. = **aecium**) and contain many **aeciospores**, which can then infect the new wheat crop.

   Label the **aeciospores**, an **aecium**, the **pycnium**, the **palisade mesophyll**, and the **spongy mesophyll** shown in Figure 9.12.

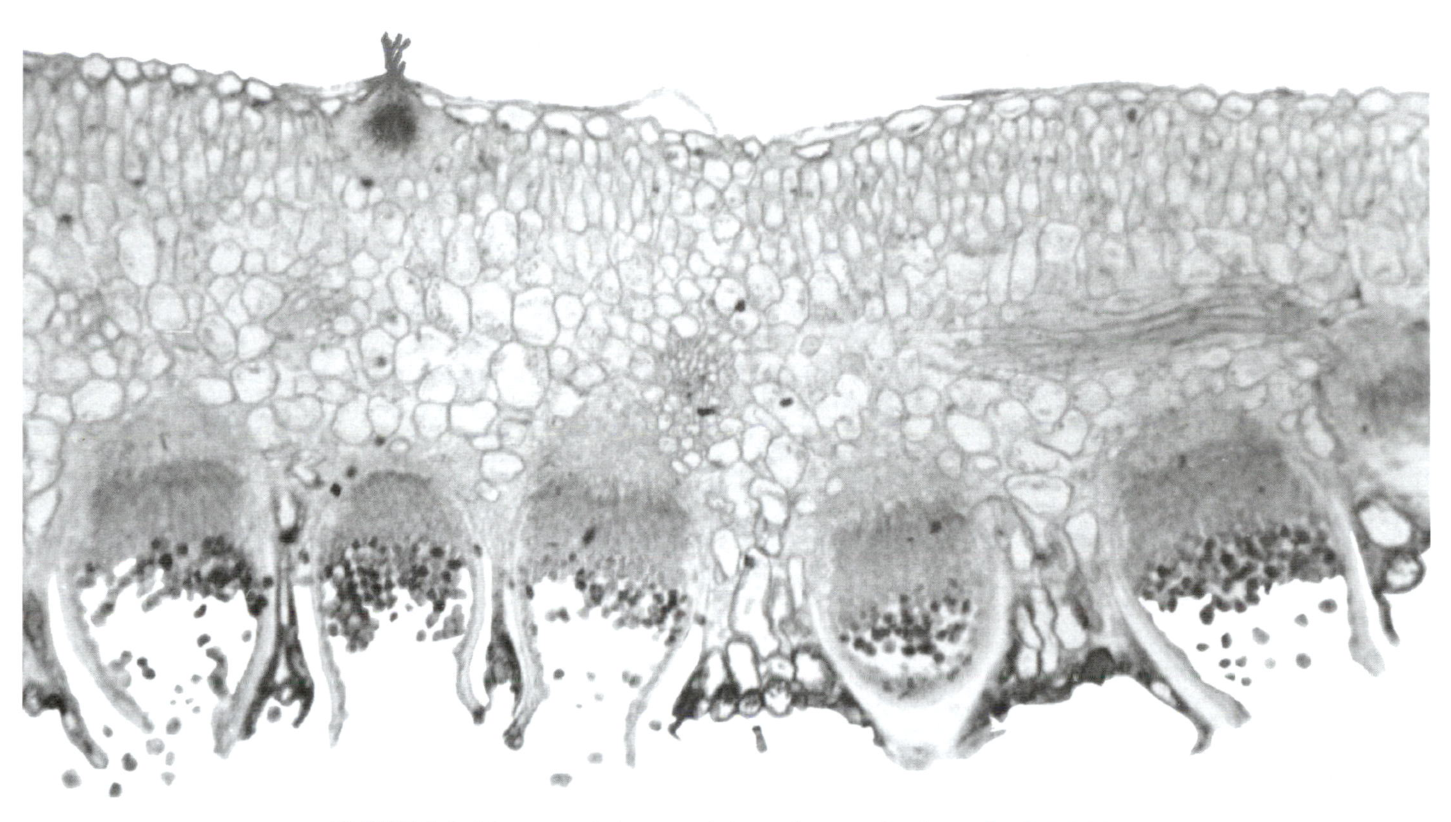

FIGURE 9.12 *Puccinia graminis* aecia on a barberry leaf, X 110.

LICHENS

Lichens are unusual plants and are difficult to classify (and identify). They are composed of two distinct parts, an alga and a fungus (most commonly, an ascomycete). The most prevalent school of thought is that these two organisms live together in a mutualistic relationship. The alga carries on photosynthesis, and the algal cells are actually penetrated by small fungal filaments, providing food for the fungus component. The fungus is thought of as providing some shelter and protection from desiccation.

A second school of thought is that this is actually a parasitic relationship; one that exists in a very delicate balance.

Even reproduction in lichens is unusual. Asexual reproduction comes about by the dispersal of small clusters of both algal cells and fungal hyphae. These structures are called *soredia*.

These two organisms are so intimately associated that the fungus cannot exist on its own. Any fungal filaments that result from spore germination must find the correct algal cells or they will die. The algae, being autotrophs, can grow independently.

Examine the demonstration slide of a longitudinal section of a lichen. Note the fungal mycelium and the algal cells that are located in a row near the surface.

There are three major growth forms found among the lichens. Small, flat lichens that are tightly appressed to a rock or other surface are said to be the **crustose** type. Lichens with a leaflike appearance are called **foliose** lichens. Lichens that are erect and have a well-developed three-dimensional shape are referred to as **fruticose** lichens.

Examine the demonstrations of these three lichen growth forms.

MYCORRHIZAE

This word translates as “fungus-root.” It is an important mutualistic relationship between the young roots of many (probably most) plants and certain fungi. The exact nature of the association is still being investigated. It seems that the fungus aids in the absorption of certain essential elements, such as phosphorus, from the soil. The plant, in return, produces sugars, amino acids, and other organic compounds for the fungus’ use.

There are two main categories of mycorrhizal associations: **ectomycorrhizae** (which are also referred to as ectotrophic mycorrhizae) and **endomycorrhizae** (endotrophic mycorrhizae). The endomycorrhizal association is much more common.

Ectomycorrhizae

In this relationship, the fungus forms a dense covering, or **mantle**, around the outside of the young root. Numerous hyphae then extend into the soil crevices, much like root hairs. Other hyphae penetrate *between* the cells of the root cortex.

Examine a prepared slide of a root section with ectomycorrhizae. Diagram a portion of the root, labeling the **mantle, root cortex,** and **penetrating hyphae** (Fig. 9.13).

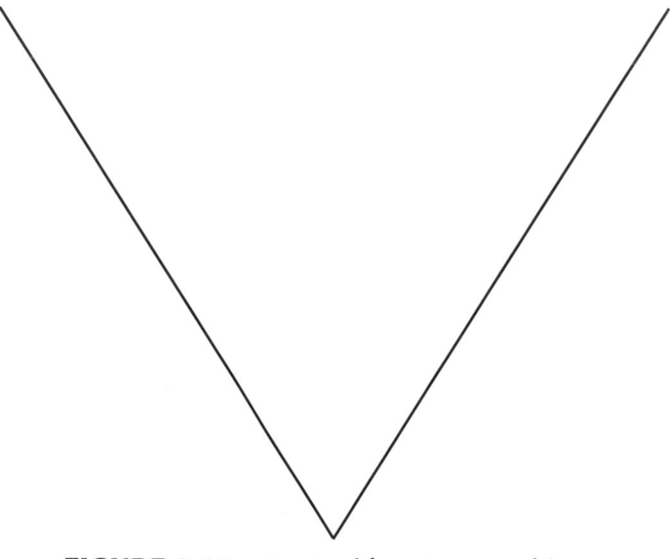

FIGURE 9.13 Root with ectomycorrhizae.

Endomycorrhizae

Here the fungal hyphae actually penetrate and can be seen *within* the cells of the root cortex. Usually there is no mantle, but some hyphae do penetrate the soil. Within the cortex cells, the hyphae form branched filaments or clusters. In some, clusters of these hyphae clump together, lose their contents, and form masses of undigested cell wall material.

Examine a prepared slide of a cross section of an orchid root with endomycorrhizae. Locate the internal hyphae, and try to locate some of the hyphae that penetrate the soil.

Make a diagram of a portion of the root, labeling the **endotrophic hyphae** and **root cortex** (Fig. 9.14).

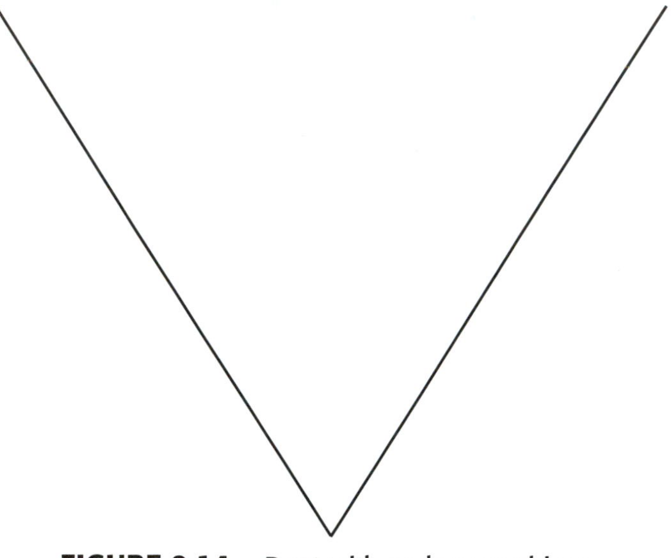

FIGURE 9.14 Root with endomycorrhizae.

Questions

1. Are ascospores haploid or diploid? (6)

2. What type of nutrition do these fungi have, i.e., parasitic or saprophytic? (7)

   Microsphaera?

   Sordaria?

   Claviceps?

   Puccinia?

3. How do the asci of *Microsphaera* and *Sordaria* differ in appearance? (8)

   Microsphaera		*Sordaria*

4. Suggest a method for controlling wheat rust without resorting to fungicides. (9)

5. How can uredospores easily be differentiated from teliospores? (10)

   Uredospores:

   Teliospores:

6. Why is it a hopeless task to try controlling mushrooms in a lawn by picking the basidiocarps? (11)

7. How does a basidium compare to an ascus in the way the spores are produced and the number of spores produced? (12)

   Basidium:

   Ascus:

8. People often have very poor success when they try to transplant native wildflowers to a home garden. Can you think of one reason for this based on information obtained in this exercise? (13)

Mosses and Liverworts

Introductory Notes

Mosses and liverworts are the simplest and most primitive of the green land plants. Commonly called Bryophytes, these plants lack true roots, stems, and leaves, as well as vascular tissue. The life cycle exhibits a well-developed ***alternation of generations*** *in which the gametophyte stage predominates.*

The reproductive structures of mosses and liverworts (and all remaining plants) are more complex than those seen in the algae and fungi. The sex organs of the algae and fungi usually consist of a single cell. In contrast, those of the bryophytes and other land plants are multicellular and have an outer protective layer of sterile jacket cells. The female sex organs, now referred to as **archegonia**, remain attached to the gametophyte plant. After fertilization, the diploid zygote develops into a multicellular embryo, another departure from the life cycles of most algae and fungi. Continued development of the embryo produces the sporophyte plant, which remains attached to the gametophtye. The sporophyte stage depends on the gametophyte for support and nourishment.

Of the three groups of plants considered in this exercise, two — the liverworts and hornworts — are inconspicuous, generally uncommon plants of little direct importance to man. Plants in the third group, the mosses, are more familiar. We will begin our study with the mosses, even though the two other groups are considered more primitive.

GOALS

After completing this exercise, the student should be able to:

- list the main characteristics of mosses and liverworts.
- list some features of these plants that may be used to classify them into separate divisions.
- explain alternation of generations as exemplified by moss plants and be able to make a diagram showing this life cycle.
- identify the parts of a moss plant from either a photograph or a diagram.
- differentiate between the three groups of bryophytes.
- identify the parts of *Marchantia* from either a photograph or a diagram.
- answer the questions in the exercise.

Because these are land plants without any vascular tissue, all of them are sometimes placed in a single division, the Bryophyta. Even though we will study only a few examples, you should see that mosses and liverworts have relatively little in common.

DIVISION BRYOPHYTA (MOSSES)

Gametophyte

The gametophyte stage is the part of the moss life cycle that is most often seen. Most are perennial and have a leafy appearance. When a moss spore germinates, it produces a filamentous structure called a **protonema** (pl. = **protonemata**). Make a wet mount of living protonema, if available, and examine these with the compound microscope.

Eventually, a bud forms on this protonema, and this will develop into the leafy gametophyte plant. Examine the prepared slides showing the development of the leafy bud. Label the **protonemata** and the **immature gametophyte** in the following photograph (Fig. 10.1).

Working with a partner, obtain a specimen of *Mnium* or *Polytrichum* which has antheridia (a male plant). Notice the leaf arrangement at the tip of the plant. At the tip of the stem is a flattened region where **antheridia** are produced. Not all mosses produce their archegonia and antheridia on separate plants. This

FIGURE 10.1 Moss protonemata with leafy shoot, X 35.

situation, as seen in these two genera, is referred to as **dioecious** or heterothallic. In many species, both male and female structures are produced on the same plant. This is a **monoecious** or homothallic condition.

Use dissecting needles to tease apart the tip of the plant and find some antheridia. Numerous, multicellular filaments can be seen in this region as well. These are **paraphyses** (sing. = **paraphysis**), and are thought to help retain a humid environment near the antheridia.

Sperm are released from the antheridia during heavy dews or rainfall. These then swim or are splashed into the vicinity of an archegonium. If you have living specimens, sperm cells may be visible in the water of your mount.

Examine a prepared slide of *Mnium* showing the antheridia and paraphyses. Note the many small cells present inside each antheridium. This is the **spermatogenous tissue**, and each of these cells will develop into a flagellated sperm cell. There is also a layer of **sterile jacket cells** surrounding each antheridium. Label the **paraphyses**, **spermatogenous tissue**, and **sterile jacket cells** in the inset of Figure 10.2.

Working with a partner, obtain a specimen of *Mnium* or *Polytrichum* that has archegonia (a female plant). Notice the leaf arrangement at the tip of the plant. The leaves are erect and enclose the stem tip. Using forceps, carefully remove the leaves until the stem tip is exposed. Cut off the stem tip, and make a longitudinal section of it. Make a wet mount and examine this with the compound microscope. Locate a vase-shaped **archegonium** (pl. = **archegonia**). The swollen basal region is the **venter**, which contains the **egg**. The elongated portion is the **neck**. The neck and venter comprise the sterile jacket cells of the archegonium. **Paraphyses** will be seen here as well and probably serve a similar function as those surrounding the antheridia.

Examine a prepared slide of *Mnium* showing the archegonia and locate an archegonbium. Label the **venter**, **egg**, **neck**, and **paraphyses** in Fig. 10.3.

FIGURE 10.3 Archegonial head of *Mnium,* X 75.

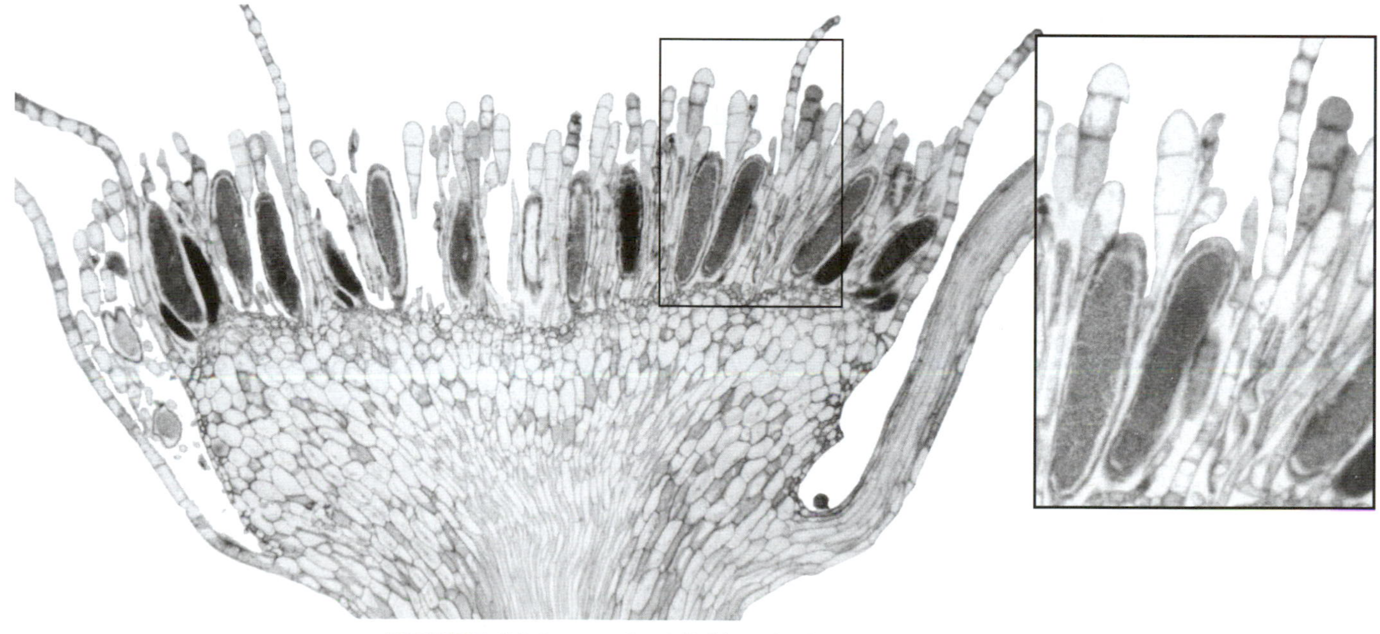

FIGURE 10.2 Antheridial head of *Mnium;* X 35, X 75.

Sporophyte

The sperms swim down a cylindrical canal in the neck of the archegonium, fertilizing the egg. The diploid zygote then develops into a sporophyte plant.

Obtain a moss plant that has an elongated structure extending from its tip. This structure is the mature sporophyte. On which plant does it develop? (1)

The **foot** is imbedded in the gametophyte tissue. The elongated part of the sporophyte is the **seta**. The swollen part at the tip is the sporangium, or **capsule**, where the **spores** are produced. A thin, papery covering may be present over the capsule. This is the **calyptra**, a remnant of the archegonium. Remove this to see the **operculum**, or lid, which is present during the maturation of the spores.

Around the inside edge of the capsule opening is a row of toothlike structures, the **peristome**. To see this, examine the capsule with the dissecting microscope. While you are watching, approach the opening of the capsule with a dropper of water. Do not touch the capsule with the water, and the capsule and peristome must be dry. The evaporating water from the dropper will bring about a change in humidity, causing the peristome to close off the opening of the capsule. When it is dry, they will open outward again. This mechanism is thought to aid in spore dispersal.

What is the advantage of having the peristome close during humid conditions? (2)

Label the **operculum, peristome, capsule,** and **seta** shown in Figure 10.4.

Examine a prepared slide of a longitudinal section of a moss capsule. Locate the central, sterile **columella, spores, operculum,** and **peristome.** Notice the large cells at the base of the operculum. These cells dry out and allow the operculum to fall off so the spores can be released. Cells such as these are often referred to as **dehiscent cells.**

Label the **columella, dehiscent cells, operculum, peristome,** and **spores** in the photograph of a moss capsule (Fig. 10.5).

FIGURE 10.4 Capsule of *Thuidium* sp., showing the operculum and peristome, X 20.

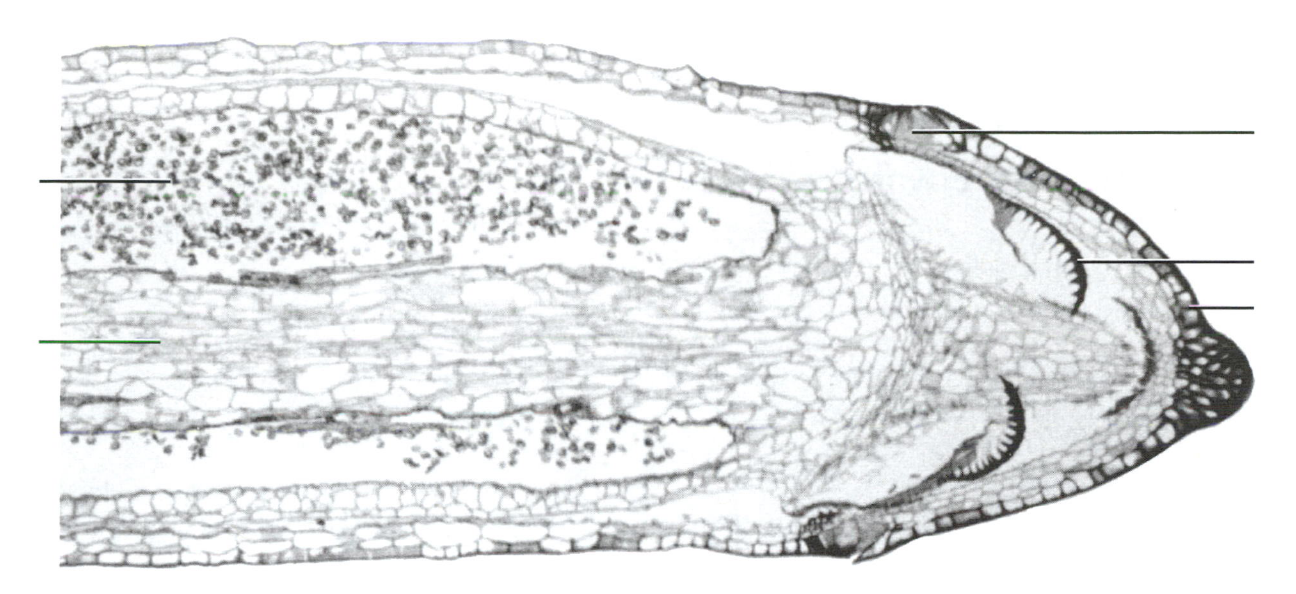

FIGURE 10.5 Longitudinal section of a moss capsule, X 35.

Sphagnum

Examine the demonstration of *Sphagnum* moss, which is commonly found in bogs and is known as peat moss. This is an important contributor to the formation of soils in boggy areas. Peat moss also has a high water-holding capacity and is often used as packing for nursery stock.

Make a wet mount of a single leaf of *Sphagnum,* and notice that there are two types of cells present. The darker cells contain chloroplasts and are referred to as **photosynthetic cells.** The larger, more numerous, clear cells, usually with conspicuous pores, are dead at maturity. These are **hyaline cells,** which retain water. Note that several photosynthetic cells form an elongated oval surrounding a number of hyaline cells.

Make a diagram of one of these oval units of a *Sphagnum* leaf, showing these two types of cells (Fig. 10.6). Label the **photosynthetic cells, hyaline cells, pores,** and **chloroplasts.**

FIGURE 10.6 Portion of *Sphagnum* leaf.

Figure 10.7 is a diagram of a moss plant showing both gametophyte and sporophyte stages. Label the diagram.

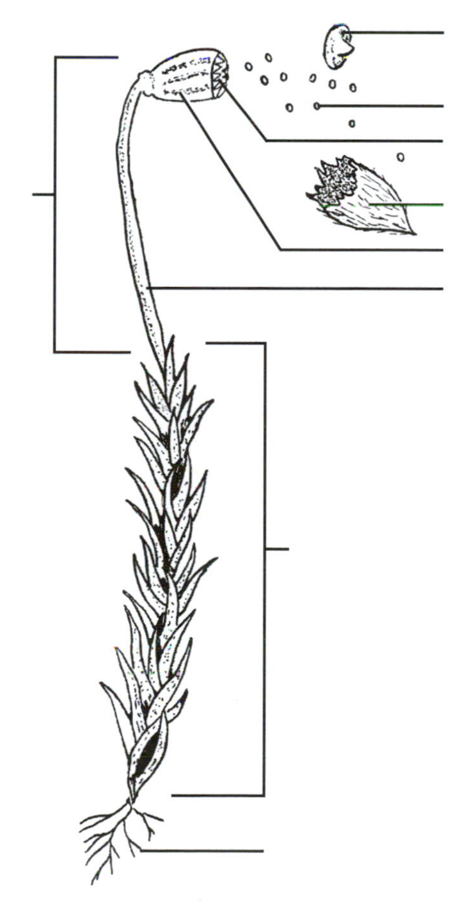

FIGURE 10.7 Moss plant with sporophyte.

DIVISION HEPATOPHYTA (LIVERWORTS AND HORNED LIVERWORTS)

Class Hepatopsida: Liverworts (Ex. *Marchantia*)

Gametophyte stage. With the aid of a dissecting microscope, examine a specimen of *Marchantia* that has **gemmae cups** present. Note the surface texturing on the dichotomously branched **thallus** (a plant body that is not differentiated into roots, stems, or leaves because there is no vascular tissue). This texturing is due to a complex system of air chambers at the upper surface. On the lower surface, you should find a number of **rhizoids**. What is their function? (3)

Examine the gemmae cup. Inside it are numerous, multicellular structures, the **gemmae** (sing. = **gemma**), which are asexual reproductive structures. Label the **thallus** and a **gemmae cup** in Figure 10.8.

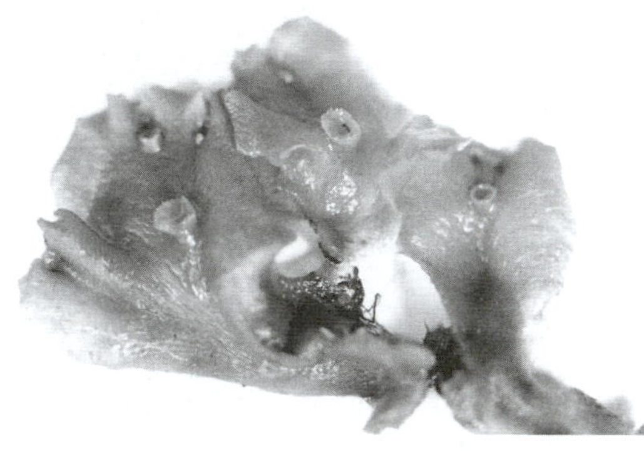

FIGURE 10.8 *Marchantia* thallus with gemmae cups, X 2.

Now examine a prepared slide of a longitudinal section of a gemmae cup. Notice the multicellular nature of the gemmae. Examine the top portion of the thallus, and notice the small photosynthetic cells and the air chambers there. The lower part of the thallus is primarily for food storage.

Label the **gemmae, rhizoids,** and **thallus** in the photograph of a *Marchantia* gemmae cup (Fig. 10.9).

Now examine some specimens of *Marchantia* plants with sexual reproductive structures on them. These structures are referred to as **antheridiophores** and **archegoniophores,** elongated structures that bear the antheridia and archegonia. Notice that the end of the antheridiophore is a relatively solid, disklike structure, while the archegoniophore terminates in a region that has several fingerlike projections. Label the **antheridiophore, archegoniophore,** and **thallus** in Figures 10.10 and 10.11.

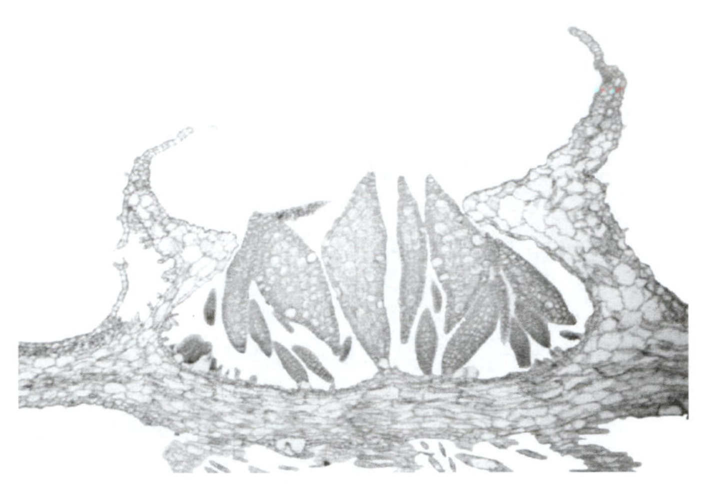

FIGURE 10.9 Longitudinal section of a *Marchantia* gemmae cup, X 50.

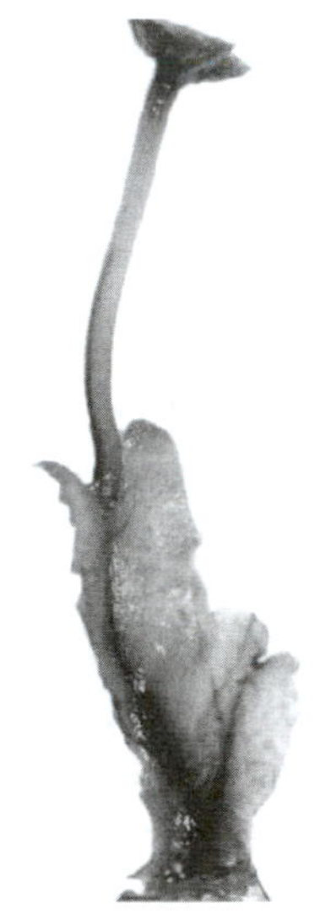

FIGURE 10.10 *Marchantia* thallus with antheridiophore, X 2.

FIGURE 10.11 *Marchantia* thallus with archegoniophore, X 2.

Examine a prepared slide of a longitudinal section of a mature antheridiophore. The **antheridia** are submerged just beneath the upper surface of the antheridiophore tip. Each antheridium has the same basic structure as those seen in the moss plant (a layer of sterile jacket cells surrounding many cells that will develop into sperm).

Label the **spermatogenous tissue, sterile jacket cells,** and **antheridiophore** in the photograph of the apical portion of an antheridiophore of *Marchantia* (Fig. 10.12).

Now examine a prepared slide of a longitudinal section of a mature archegoniophore with **archegonia.** The archegonia are flask-shaped and located on the lower surface of the arms of the archegoniophore. Their general structure is similar to that of the moss archegonia.

Figure 10.13 shows the general location of the *Marchantia* archegonia.

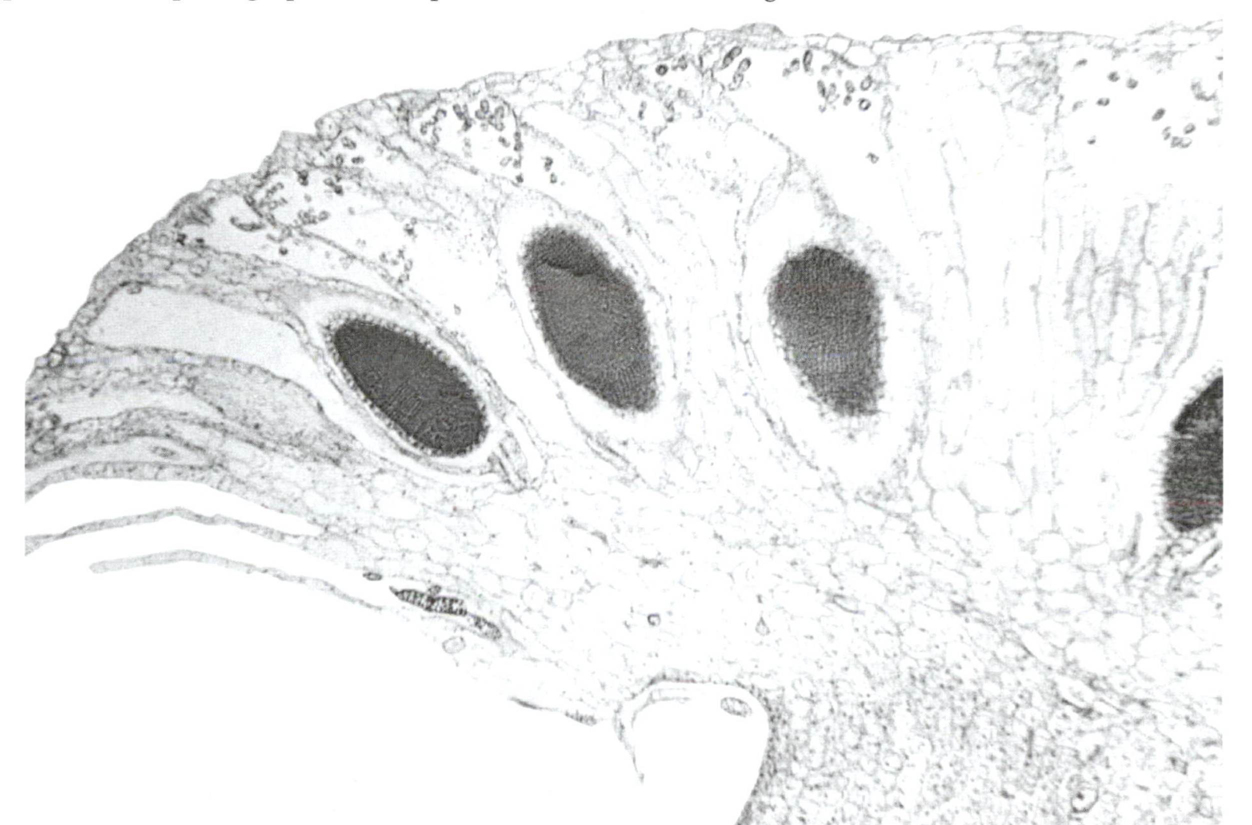

FIGURE 10.12 *Marchantia* antheridia, X 80.

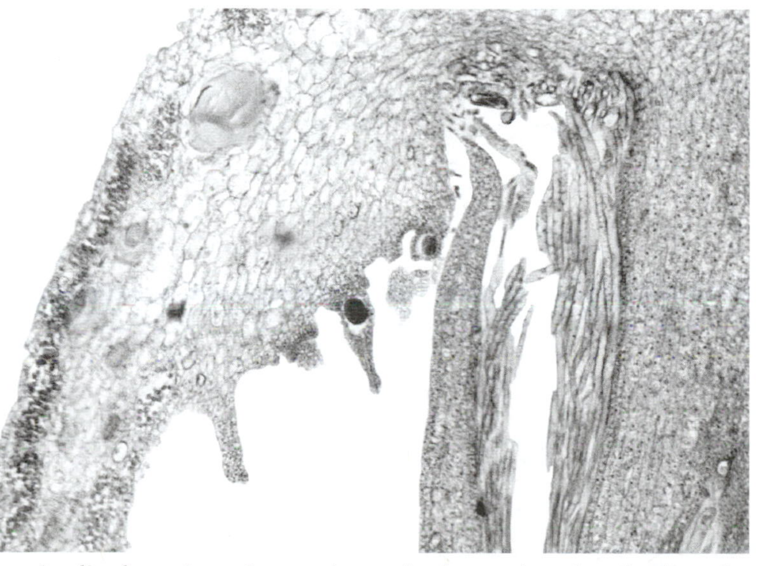

FIGURE 10.13 Longitudinal section of an archegoniophore, showing the location of the archegonia, X 50.

Use Figure 10.14 to label the **egg**, **venter**, **neck**, and **archegoniophore** of the *Marchantia* archegonium and archegoniophore.

Sporophyte stage. After fertilization, the zygote develops into the diploid sporophyte. Figure 10.15 shows a *Marchantia* plant with fully developed sporophytes. Note that these sporophytes are in the same location as the archegonia.

Label the **sporophytes**, **gametophyte thallus**, and **archegoniophore** as seen in Figure 10.15.

Examine a prepared slide of a *Marchantia* sporophyte. The sporophytes of *Marchantia* are much smaller and less conspicuous than those of the mosses.

Several stages of development may be seen. Young sporophytes show little cellular differentiation. In more mature sporophytes, a basal **foot**, a midregion or **seta**, and a swollen, terminal **sporangium** can be seen.

Within the sporangium are many **spores** and elongated threadlike structures called **elaters**. Examine these elaters, and note the spiral or braidlike nature of the cellulose walls. These elaters respond to changes in humidity by coiling and uncoiling within the sporangium.

FIGURE 10.15 *Marchantia* archegoniophore with sporophytes on the lower surface, X 2.

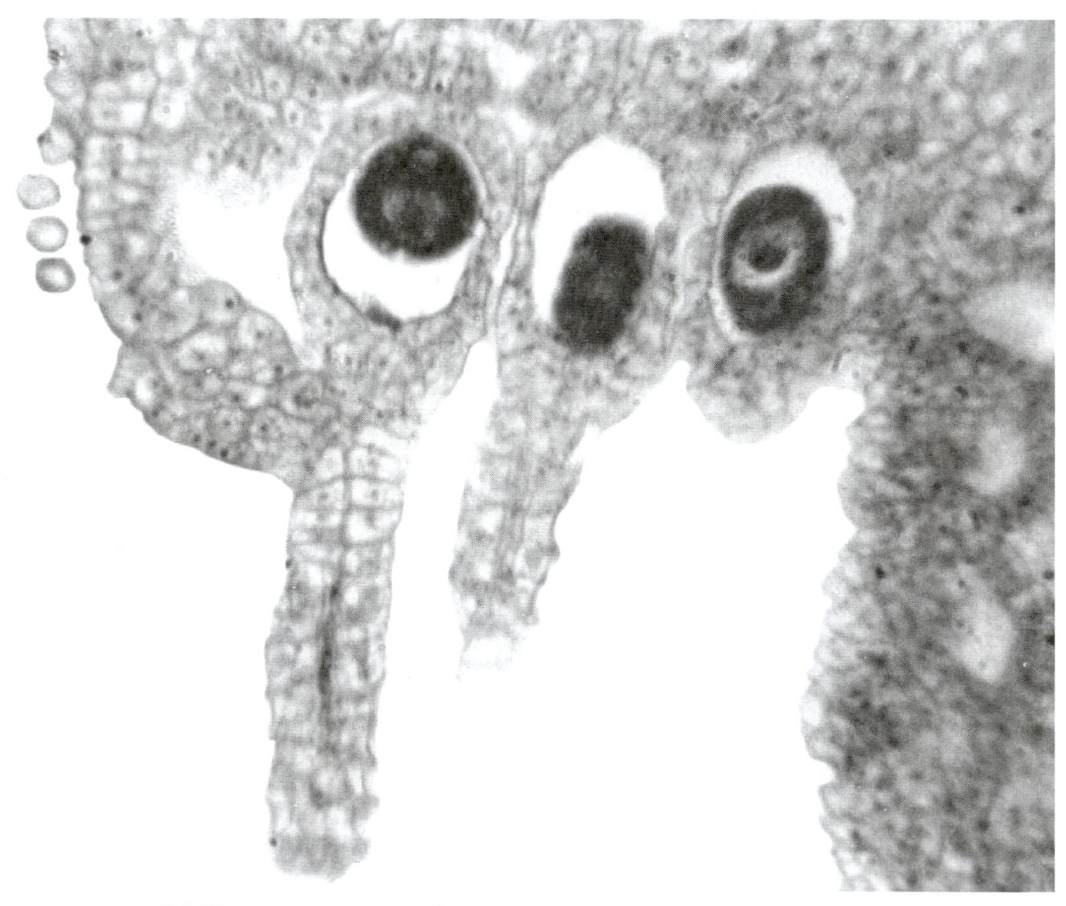

FIGURE 10.14 *Marchantia* archegonia, X 390.

The sporangium may be partially enclosed by the remains of the archegonium, the **calyptra,** as in the mosses.

Look for tetrads of spores.

Are the spores haploid or diploid? The elaters? (4)

What is the function of the elaters? (5)

Label the **calyptra, foot, seta,** and **sporangium with spores** in the mature sporophtyes (Fig. 10.16).

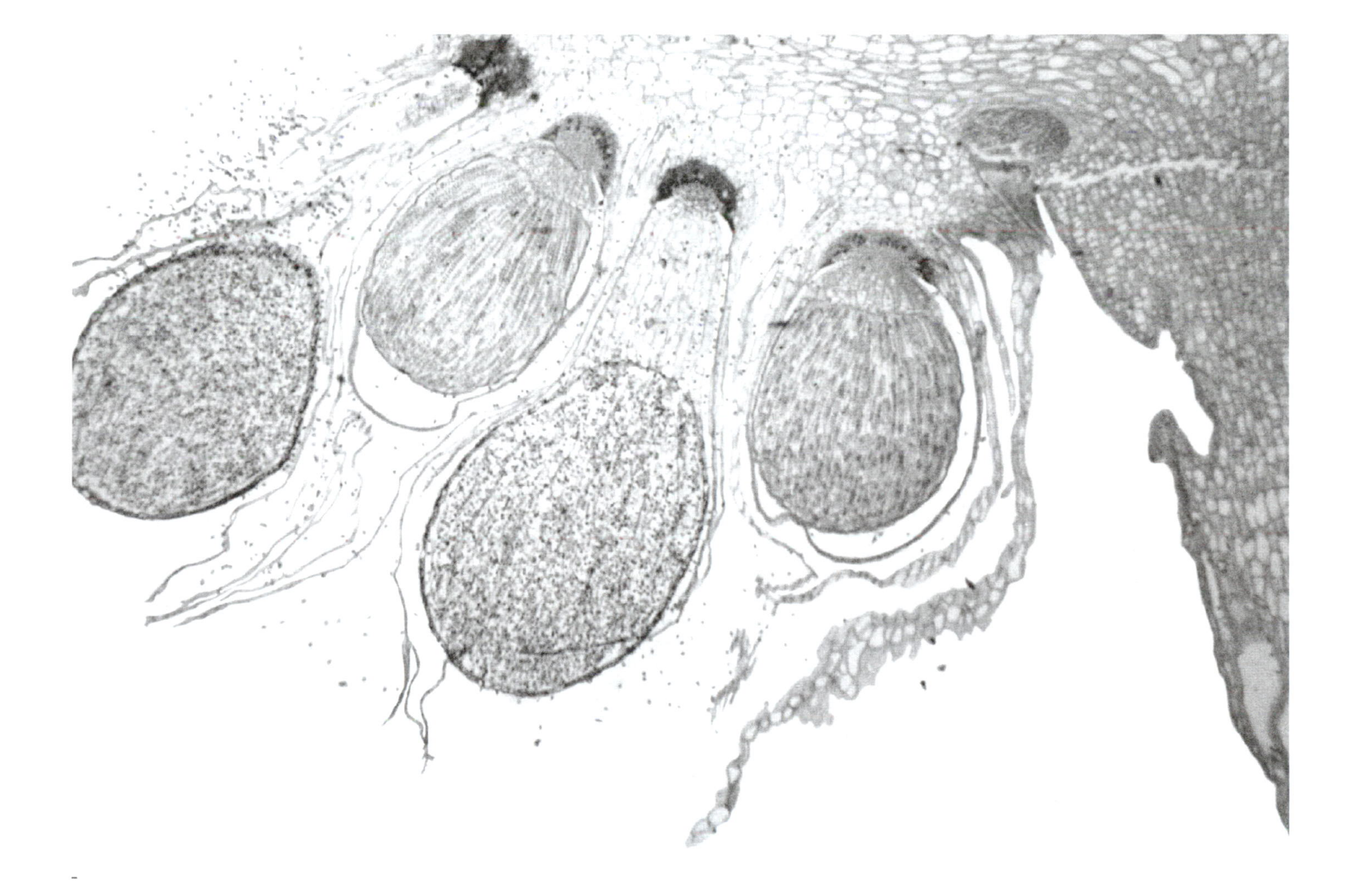

FIGURE 10.16 *Marchantia* sporophytes, X 75.

Class Antheroceropsida: Horned Liverworts (Ex. *Anthoceros*)

Examine a prepared slide of *Anthoceros*. These specimens are gametophyte plants with sporophytes attached. The elongated appearance of the sporophyte is responsible for the name of the group. Archegonia and antheridia are imbedded within the **gametophyte thallus.** The **sporophyte** that develops is an elongated, spikelike structure. The **foot** is imbedded in the gametophyte thallus. Just above the foot is a region of meristematic cells. This **basal meristem** gives rise to all of the cells found in the **sporangium** (capsule) above it.

Label the **gametophyte thallus, foot, basal meristem,** and **sporangium with spores** in the photograph of *Anthoceros* (Fig. 10.17).

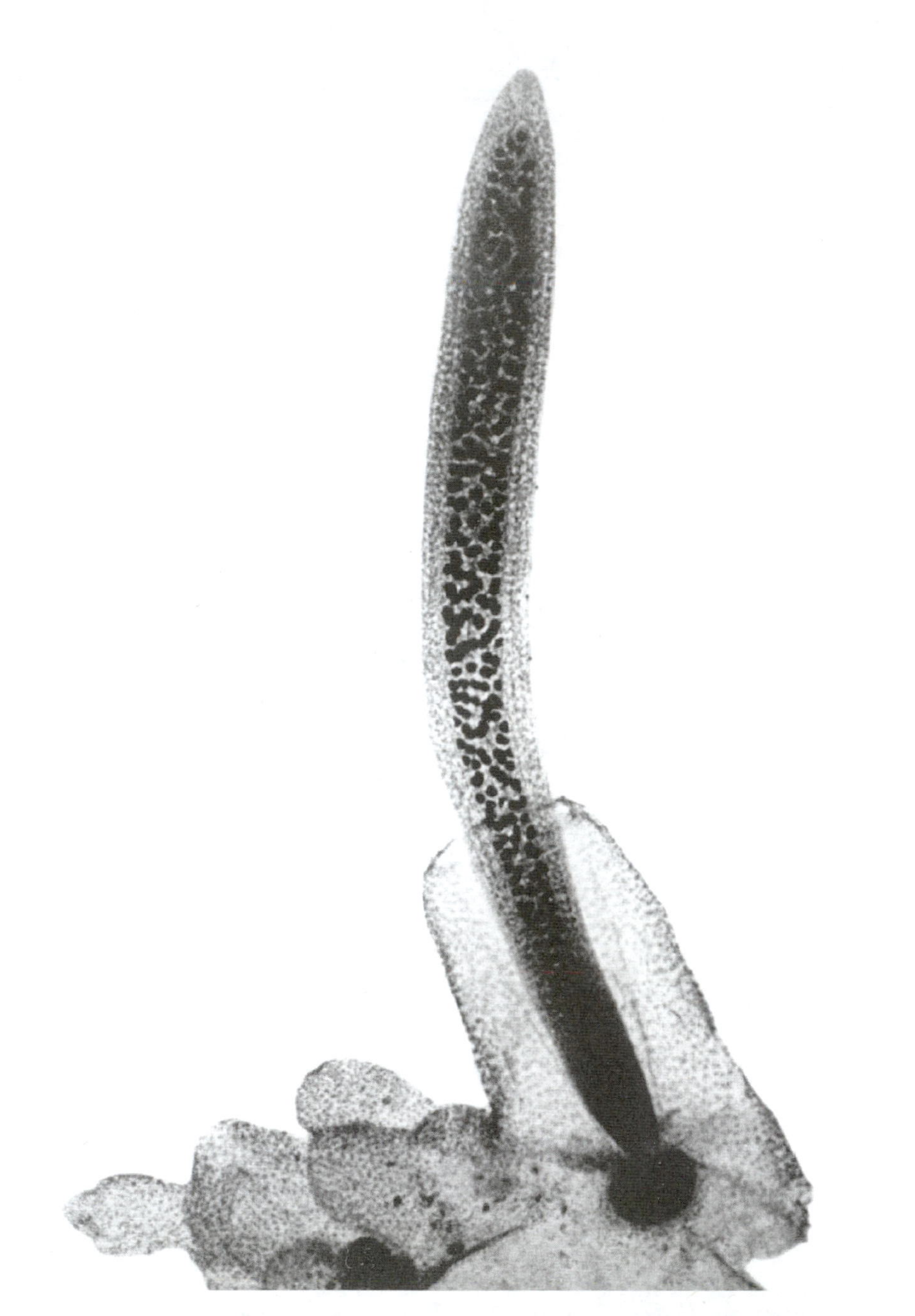

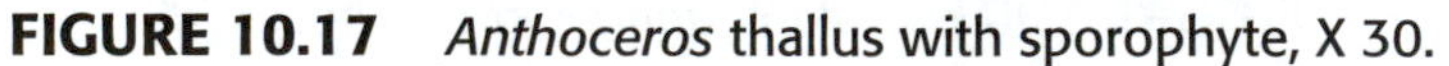

FIGURE 10.17 *Anthoceros* thallus with sporophyte, X 30.

Questions

1. What is meant by alternation of generations? (6)

2. What type of *nuclear division* is involved in the development of a sporophyte plant? Of the spores? (7)

3. Arrange the following structures or processes into a diagram showing the life cycle of a moss plant. (8)

Antheridium	Mature female plant	Spores
Archegonium	Mature male plant	Spore dispersal
Capsule	Mature sporophyte	Spore germination
Egg	Meiosis	Spore mother cells
Embryo	Protonema	Tetrad of spores
Fertilization	Sperms	Young sporophyte
Leafy bud	Sperm transfer	Zygote

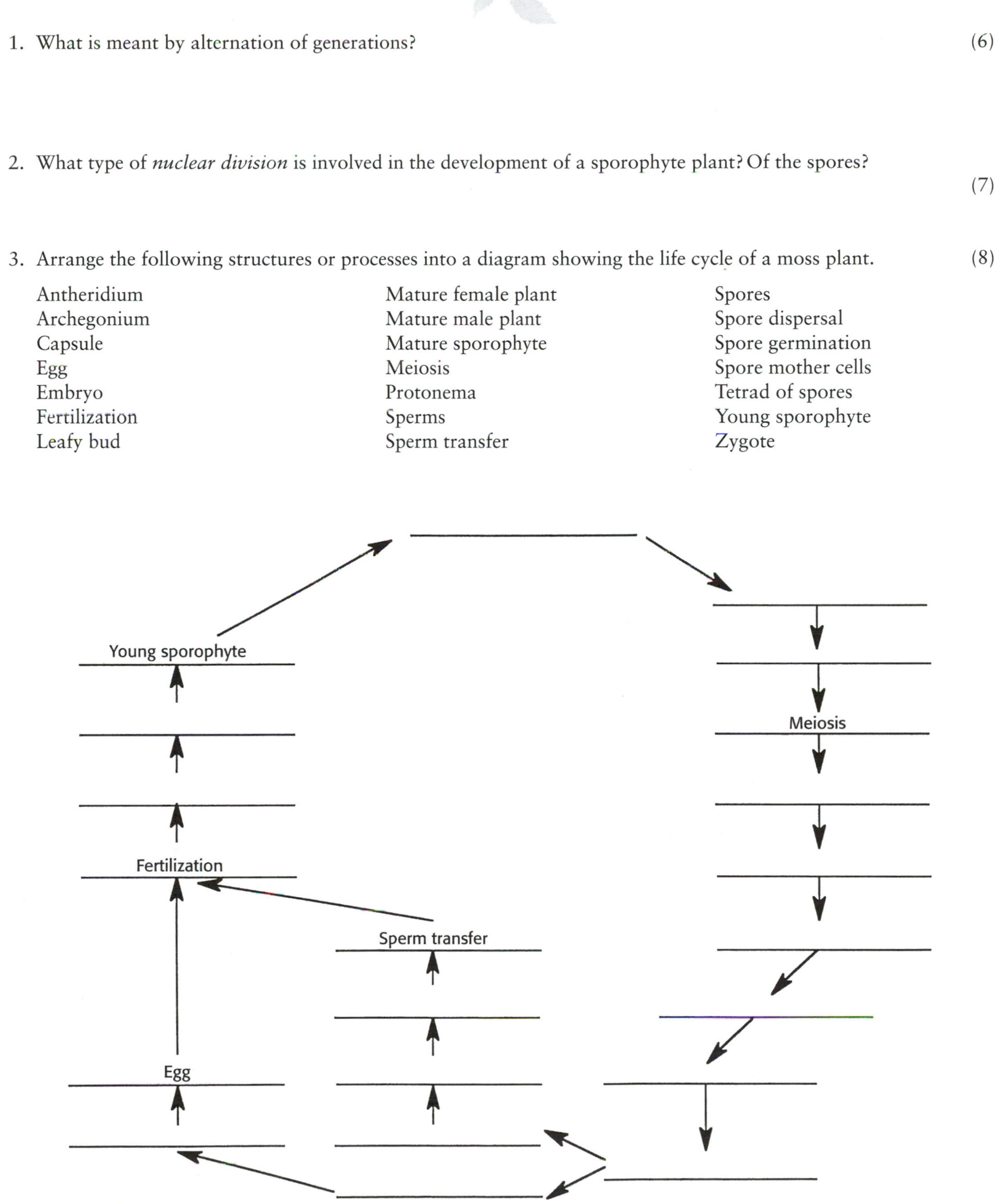

4. How are the gemmae of *Marchantia* dispersed? (9)

5. The leaves, stems, and rhizoids of mosses are analogous to leaves, stems, and roots of higher plants. What does this mean? (10)

6. Compare the relative dominance of the gametophyte versus the sporophyte of the mosses and liverworts. (11)

Primitive Vascular Plants

Introductory Notes

Vascular plants are characterized by the presence of specialized cells for conducting water and food materials throughout the plant. These tissues are the ***xylem****, which is responsible for conducting water and dissolved soil nutrients upward, and the* ***phloem****, which conducts sugars in solution. These vascular tissues are present only in the sporophyte plant. As a result, the sporophyte plant is much larger than, and independent of, the gametophyte (in contrast to the bryophytes). Gametophytes will not be studied.*

DIVISION PSILOPSIDA (PSILOPHYTES)

Members of this plant division are very primitive vascular plants with a fossil record extending back some 300–400 million years. Most psilophytes are known only as fossils, with only two genera extant. *Tmesipteris* is found in the South Pacific region, but *Psilotum* is more widespread in tropical parts of the world (it grows in Florida and westward into Texas). Both genera may grow as true terrestrial plants or as epiphytes.

Psilotum nudum

Vegetative Structures

1. Examine the living and/or preserved specimens of *Psilotum*. Notice that it has a very simple form. There is a horizontal stem, or **rhizome**, which is usually beneath the soil surface. No true roots are present, but there are rhizoids. Both the rhizome and rhizoids function for anchorage and absorption. *Psilotum* also lacks true leaves. The small leaflike structures seen along the stem do not have any vascular tissue and are known as **enations**. Nearly all of the photosynthesis occurs in the chlorenchyma tissue found in the stem, which exhibits a primitive

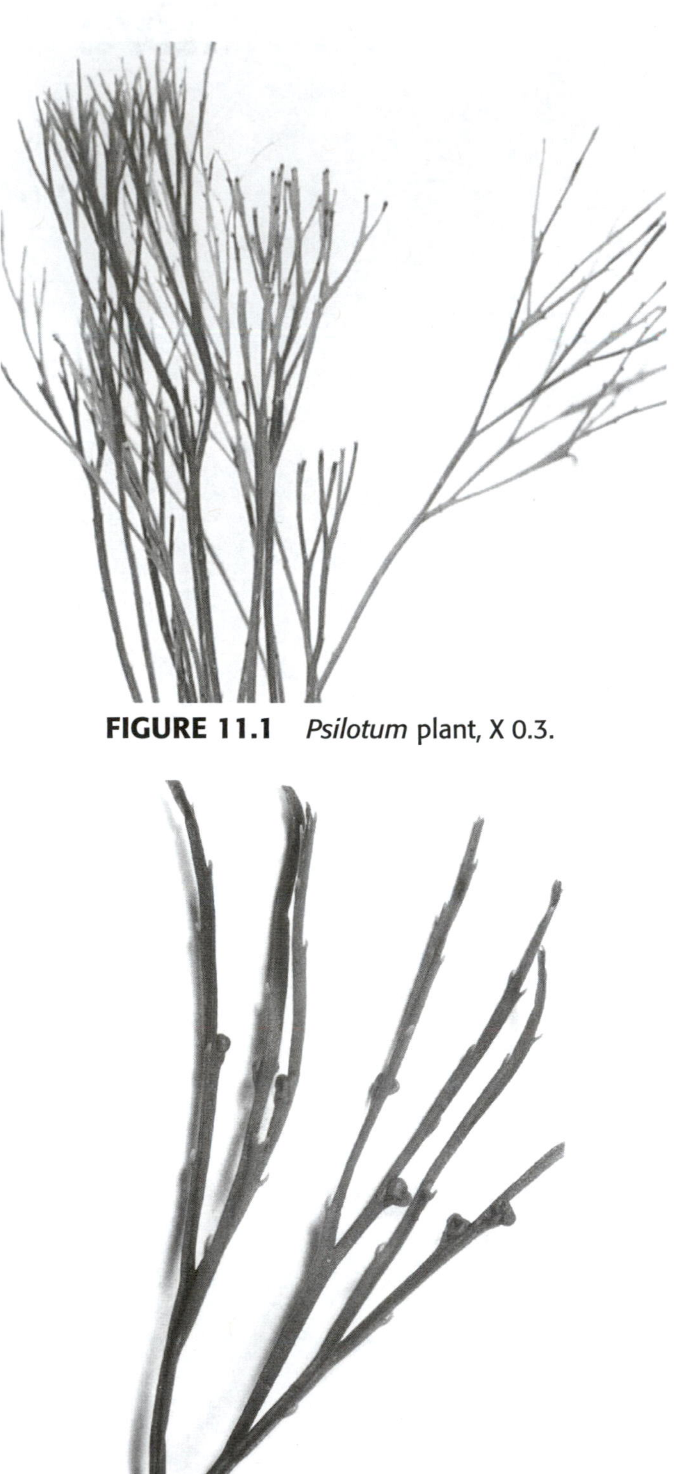

FIGURE 11.1 *Psilotum* plant, X 0.3.

FIGURE 11.2 *Psilotum* branches with sporangia, X 1.

GOALS

After completing this exercise, the student should be able to:

- list several differences between vascular plants and bryophytes.
- explain why the Psilopsida are considered the most primitive vascular plants.
- identify, from either a photograph or a diagram, the four genera studied and various structures of the plants examined.
- answer the questions in the exercise.

type of branching known as **dichotomous branching.** Spherical sporangia are also produced along the stem. Typically, three sporangia fuse together, forming what is known as a **synangium** (pl. = **synangia**). *Psilotum* is commonly known as Whisk Fern.

2. Now examine a prepared slide of a *Psilotum* stem cross section. Notice the **epidermis** with its well-developed **cuticle. Stomata** are present here as well. The **cortex** is relatively large and has three regions. Beneath the epidermis is a region of parenchyma cells that contain chloroplasts (**chlorenchyma**). Adjacent to the chlorenchyma is typically a zone of **sclerenchyma** cells that give the stem support. The innermost region of the cortex consists of another region of parenchyma tissue, modified for food storage. In the very center of the stem is the single bundle of vascular tissue. The **xylem** is found in the very center and is surrounded by the **phloem.** This type of stele is referred to as a **protostele.** At the junction between the stele and the cortex is an **endodermis.** A **Casparian strip** should be easily seen here.

3. Label the **cuticle, epidermis, guard cells and stoma, chlorenchyma tissue, food storage region of the cortex, xylem,** and **phloem** seen in Figure 11.3.

Reproductive Cycle

Examine a prepared slide of a cross section of a *Psilotum* synangium. Here the three-part nature can be seen easily. Within each compartment, diploid spore mother cells divide by meiosis forming four haploid spores. The spores are oval structures with a netted surface texture. Locate the **dehiscent cells,** several specialized cells in the outer wall of each sporangium that dry up when the spores are mature. The drying of these cells allows the sporangia to split open, releasing the mature spores.

Label the **dehiscent cells** and **spores** as seen in Figure 11.4.

Spore germination is quite slow, but eventually, a haploid gametophyte plant is produced. In *Psilotum,* the gametophyte is roughly cylindrical and may be only several millimeters in length. The gametophyte is monoecious, having both antheridia and archegonia present. It is usually subterranean and has endomycorrhizal fungi.

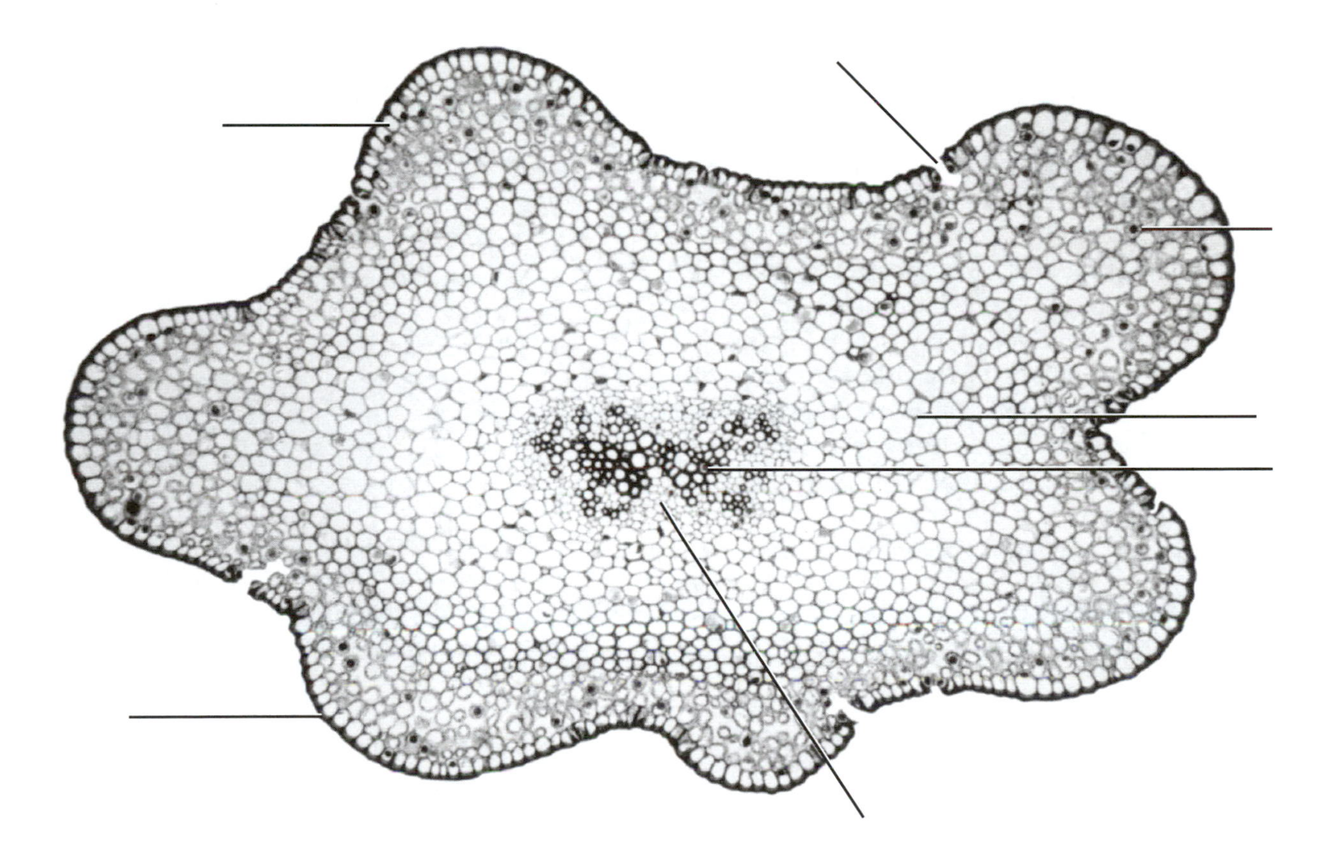

FIGURE 11.3 *Psilotum*, cross section of an aerial stem, X 95.

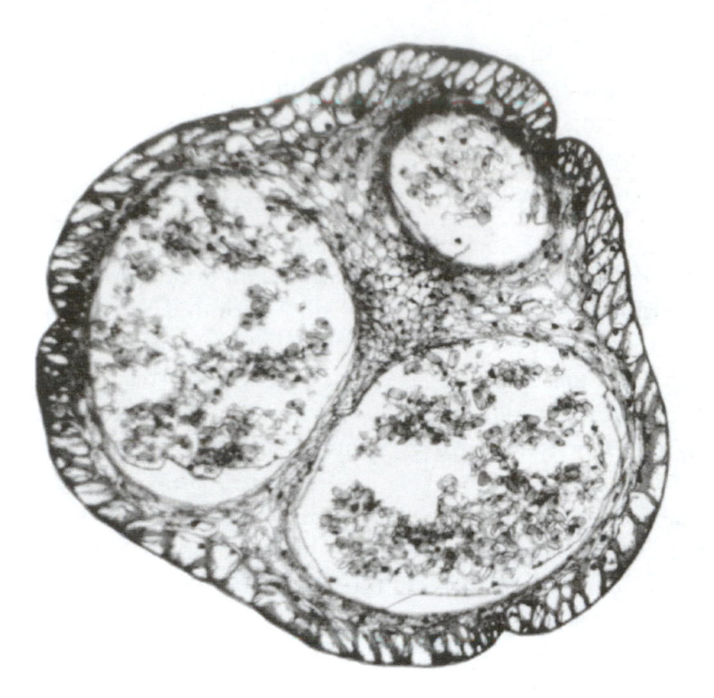

FIGURE 11.4 *Psilotum*, cross section of a synangium, X 45.

DIVISION MICROPHYLLOPHYTA (CLUB MOSSES)

The club mosses and their allies were formerly much more abundant than they are now. Present day plants are inconspicuous, small plants. Some of the extinct forms were quite large, over 100 feet tall. During the Carboniferous period, they formed large forests in many parts of the world. Our present fossil fuel deposits are the result of the slow decomposition of these plants. They are probably just as ancient as the psilophytes.

The club mosses are seedless vascular plants that differ from the psilophytes in that they *do have true roots and leaves*. The small leaves are supplied with vascular tissue, but there is no leaf gap present in the stem where the leaf trace develops. Leaves such as these are called **microphylls**.

The two largest and most common genera of the Microphyllophyta are *Lycopodium* and *Selaginella*. Three other, less common, genera are recognized. *Lycopodium* is the most common genus in our geographic region.

Lycopodium

1. Examine the herbarium mounts of several local species and then obtain a preserved plant.

   a. List two differences in outward appearance between *Lycopodium* and *Psilotum*. (1)

   b. How are these two genera somewhat similar in appearance? (2)

   c. Why do you suppose they are called club mosses? (3)

2. The sporangia of *Lycopodium* are associated with modified leaves called **sporophylls**. In some species, such as *L. lucidulum*, most of the leaves along the stem are able to produce sporangia. In most, however, the sporophylls are clustered into conelike structures called a **strobili** (sing. = **strobilus**) at the tips of the stems. What is a possible advantage of having this location for the strobili? (4)

The spores of *Lycopodium*, which in some species are produced in great abundance, have been put to a variety of peculiar uses. Some of these include their inclusion in fireworks, a dusting on pills to keep them from sticking together, and, in the early days of photography, as a flash powder.

Remove one or two sporophylls, *with the sporangium attached*, from the strobilus. Examine them with the dissecting microscope. Is the sporangium located on the upper or lower surface of the sporophyll? (5)

Make a labeled diagram showing a **sporangium** that is *still attached* to an intact **sporophyll** (Fig. 11.5)

FIGURE 11.5 *Lycopodium* sporophyll with sporangium.

Crush several sporangia. Notice that all of the spores are the same size.

If there is a single kind of spore produced, the plant is **homosporous**. If there are two kinds of spore produced, one larger than the other, the plant is said to be **heterosporous**.

Lycopodium is homosporous. Confirm this by examining a prepared slide of a longitudinal section of a *Lycopodium* strobilus.

Label a **sporophyll**, **sporangium**, and **spores** in the photograph of a longitudinal section of a *Lycopodium* strobilus (Fig. 11.6).

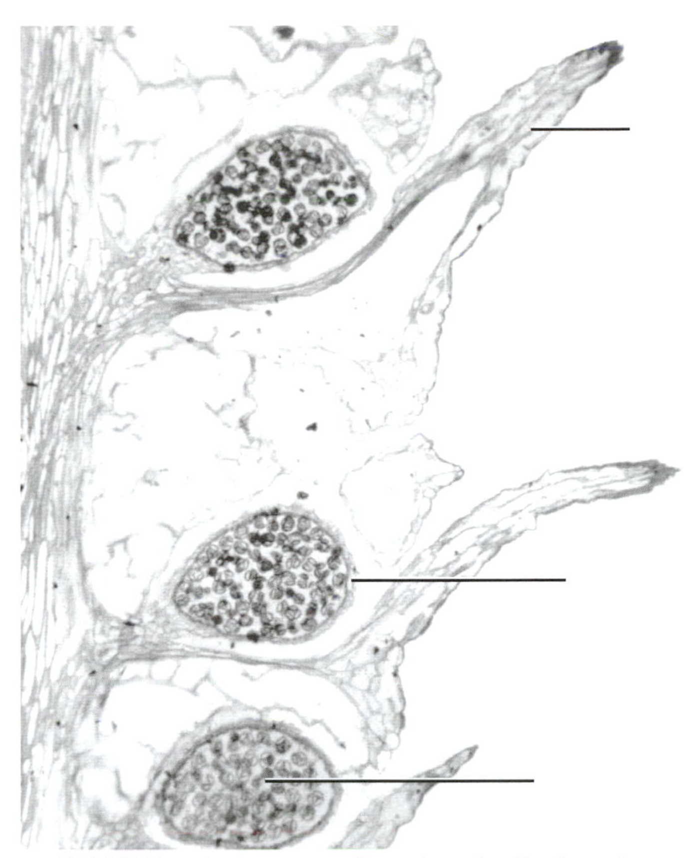

FIGURE 11.6 *Lycopodium*, longitudinal section of a strobilus, X 30.

Selaginella

The genus *Selaginella* contains approximately 700 species, most of which are tropical. While those species found in temperate zones are small and inconspicuous, some of the tropical forms attain much greater size.

Obtain a preserved specimen of *Selaginella*. Notice how it differs in appearance from *Lycopodium*.

Selaginella is heterosporous, and all species bear their sporophylls in strobili. Use the dissecting microscope to examine the strobilus of your specimen, and look for sporangia at the base of the sporophylls. You should detect two kinds, which differ in size, shape, and color.

1. Select two sporangia, *one of each kind*, and place them side by side for comparison. One of the sporangia should be larger than the other and should also have some conspicuous bulges. *Selaginella* is heterosporous, having larger **megaspores** and smaller **microspores**. The sporangium with the larger spores is called a **megasporangium**. The other is a **microsporangium**. The sporophylls are also named accordingly (**megasporophyll** and **microsporophyll**).

2. Using a dissecting needle, carefully rupture the sporangia to release the megaspores and microspores.
3. Make two labeled diagrams showing the open mega- and microsporangia and associated spores (Figs. 11.7 and 11.8). Label the diagrams appropriately, i.e., **megasporangium, megaspore, microsporangium** and **microspores.**

FIGURE 11.7 *Selaginella*, megasporangium and megaspores.

FIGURE 11.8 *Selaginella*, microsporangium and microspores.

a. How many microspores are present? (6)

b. How many megaspores? (7)

4. Examine a prepared slide of a longitudinal section of a *Selaginella* strobilus.
5. Label a **microsporophyll,** a **megasporophyll,** a **microsporangium,** a **megasporangium, microspores,** and a **megaspore** in Figure 11.9.

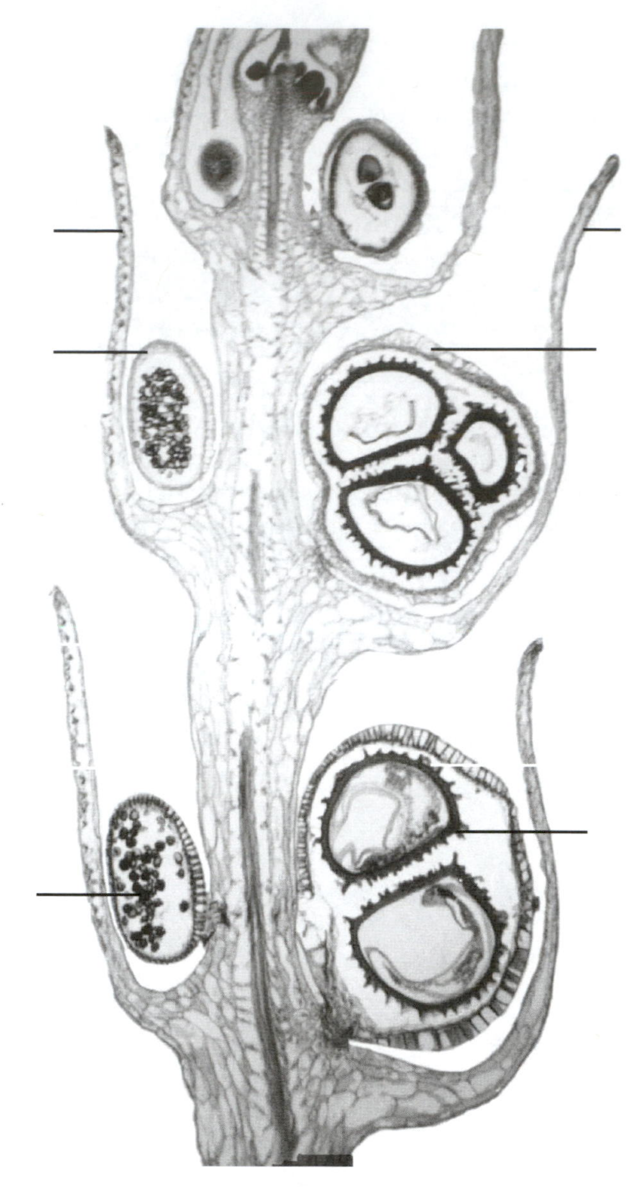

FIGURE 11.9 *Selaginella*, longitudinal section of a strobilus, X 40.

DIVISION ARTHROPHYTA (HORSETAILS)

Equisetum

The horsetails represent another group of plants with an extensive Devonian fossil record. Like the club mosses, these plants once formed an important assemblage of land plants. The only genus still extant is *Equisetum*.

Morphology

Examine preserved plants and herbarium specimens of *Equisetum*. Some species have two distinctly different sporophyte stages: a fertile stage that produces strobili (Fig. 11.10), and a sterile stage that is green and photosynthetic (Fig. 11.11). Examine the sterile stage, and notice the many **photosynthetic branches** and **scale leaves** that arise at the **nodes** (joints). The portion of the stem between two adjacent nodes is the **internode**.

The leaves are small, inconspicuous, scalelike structures produced in a whorl at each node.

How does the general appearance of *Equisetum* differ from that of *Lycopodium*? (Compare a fertile shoot with one of the herbarium mounts of *Lycopodium*.) (8)

The strobilus is similar to that of the club mosses, with a few modifications. The sporophyll is highly modified and is referred to as a **sporangiophore**. In the *Equisetum* strobilus, each of the hexagonal units is a sporangiophore.

Figures 11.10 and 11.11 show both stages of *Equisetum*. Label the **strobilus, sporangiophore, node, internode,** and **scale leaf** in Figure 11.10.

Anatomy

Examine a prepared slide of an *Equisetum* stem cross section. This is a section through the internodal region. Note the conspicuous ridges on live or preserved specimens. The **epidermis** is composed of thick-walled cells, and **stomata** may be found here. The stem of the plant incorporates large deposits of silica, giving an alternate name — scouring rush.

Three distinct canals can be found. The hollow pith is commonly referred to as the **central canal**. The ring of small canals just to the outside of the central canal and opposite the ridges are the **carinal canals**. These mark the location of the vascular bundles and may also function in conduction as well. The third type of canal are the medium-sized **vallecular canals**, which lie opposite the grooves in the stem.

The **vascular tissues** can be found just to the outside of the carinal canals. These vascular bundles consist of primary phloem, bordered on each side by a few primary xylem cells. An endodermis separates the vascular tissue from the cortex.

Label *each* of the **three canals**, the **vascular tissue**, and a **stoma** in the portion of an *Equisetum* stem shown in Figure 11.12.

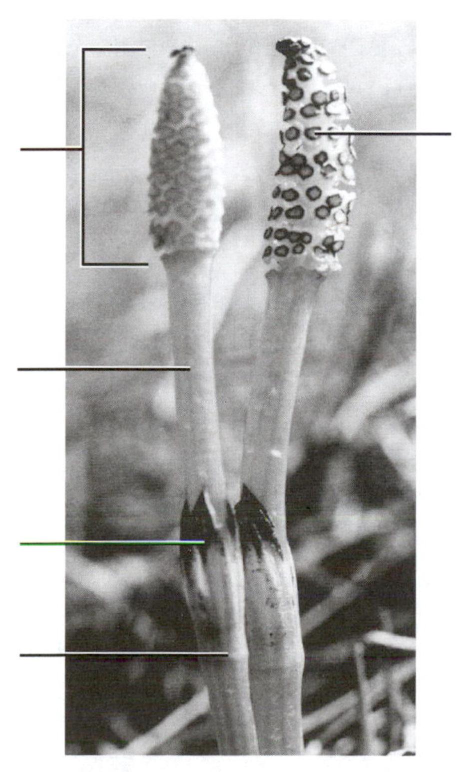

FIGURE 11.10 Fertile shoot of *Equisetum*, X 1.

FIGURE 11.11 Sterile shoot of *Equisetum*, X 0.3.

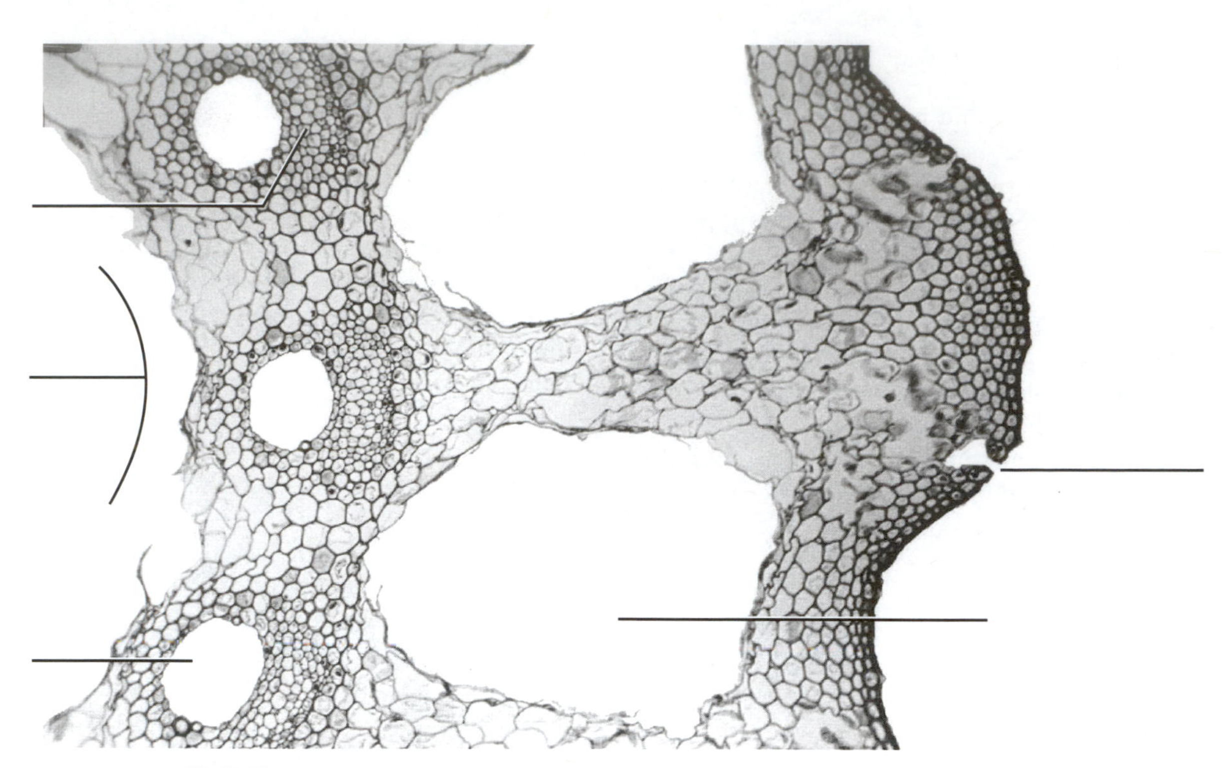

FIGURE 11.12 Cross section of an aerial stem of *Equisetum,* X 110.

Asexual Reproduction

Obtain a preserved or fresh *Equisetum* strobilus. Remove a sporangiophore and examine it with the dissecting microscope. Notice the general umbrella shape and the sporangia borne on the lower surface.

Make a labeled diagram of an *intact* **sporangiophore** showing the **sporangia** (Fig. 11.13).

FIGURE 11.13 *Equisetum* sporangiophore with sporangia.

Crush a sporangium on a clean, dry slide. Do not add water or a cover slip. Examine the spores with the low power of the compound microscope. Notice that each spore has a pair of elongated appendages called **elaters.**

These elaters are hygroscopic (respond to changes in humidity). Notice how they react as the spores dry out. When the spores are dry, carefully exhale on them to see how they respond to a more humid situation.

How might this aid in spore dispersal? (9)
(Consider both dry and humid conditions.)

Make a labeled diagram of some *Equisetum* **spores** with **elaters** (Fig. 11.14).

FIGURE 11.14 *Equisetum* spores with elaters.

Examine a prepared slide of a longitudinal section of an *Equisetum* strobilus. Find the **main axis** of the strobilus, a place where a **sporangiophore** is attached, a **sporangium,** and **spores with elaters**. Label these structures on Figure 11.15.

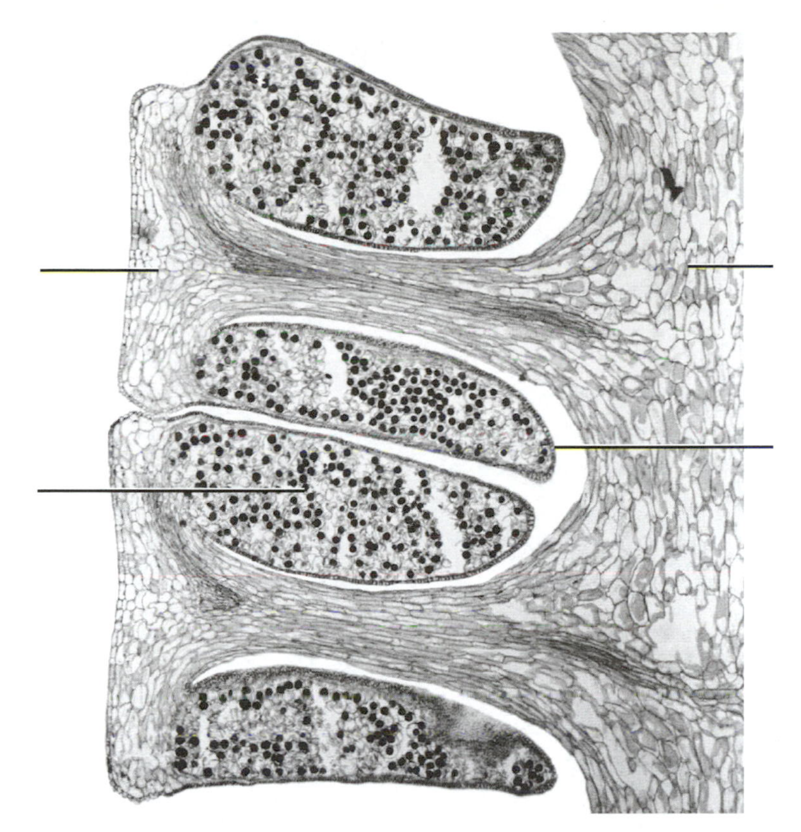

FIGURE 11.15 Longitudinal section of *Equisetum* sporangiophores, X 50.

Questions

1. The plants examined in this exercise are considerably larger than the bryophytes. Why? (10)

2. What are two characteristics of *Psilotum* which indicate that it lives in dry habitats? (11)

3. What are two traits of *Psilotum* that make it the most primitive of the vascular plants? (12)

4. How does the life cycle of these plants differ from that of the bryophytes? (13)

   Bryophytes:

   Plants from this lab:

5. Define or explain the following terms as they relate to this lab: (14)

Rhizome:

Sporophyll:

Equisetum:

Microsporophyll:

Megaspore:

Strobilus:

Protostele:

Elaters:

Heterosporous:

6. Complete the following table: (15)

Feature	***Psilotum***	***Lycopodium***	***Selaginella***	***Equisetum***
Dichotomous branching?			Yes	
Leaves		Microphylls		
Roots				True roots
Strobilus present?				
Homosporous or Heterosporous				

Division Pterophyta: Ferns

Introductory Notes

*Ferns are the largest and most important group of seedless vascular plants. They differ from the other primitive land plants because their leaves are **megaphylls**. This leaf has a more complex venation and also exhibits a **leaf gap** in the stem. There is a well-developed alternation of generations in which both the sporophyte and the gametophyte are independent at maturity. The sporophyte stage of the fern is the most familiar, with most people never seeing or even knowing about the gametophyte.*

GOALS

After completing this exercise, the student should be able to:

- describe and diagram the main vegetative structures of a fern sporophyte.
- describe the life cycle of a fern.
- list several differences between ferns and mosses.
- recognize, from either photographs or diagrams, the various structures studied in the exercise.
- answer the questions in the exercise.

SPOROPHYTE

External Morphology

Examine the demonstrations of fern plants in the room. The part of the plant that is visible above the soil level is the leaf, or **frond**. Typically, the stem is represented by a horizontal, underground **rhizome**. True roots are also present.

Examine a fern frond. Note that it is composed of two main parts: the basal **petiole** and the flattened, expanded **blade**. In most ferns, the blade is subdivided and has a central axis. This central axis, or **rachis,** is an extension of the petiole. Each subdivision of the blade, extending from the rachis, is called a **pinna** (pl. = **pinnae**). These are similar to the leaflets of a pinnately compound leaf. In some species, the pinnae themselves are divided into **pinnules**. The complex nature of the fern leaf is why they are so often grown as house plants.

Herbarium mounts are available to show these features and to illustrate the rhizome and roots.

Figure 12.1 is a photograph of *Polypodium vulgare*, a fern common to much of the eastern United States where there are extensive rocky ledges. Label the **roots, rhizome, petiole, blade, frond,** and a **pinna**.

Internal Anatomy

1. Use a dissecting microscope to examine a prepared slide of the rhizome of a maidenhair fern, *Adiantum*. This is an excellent example of a **siphonostele** (the vascular tissue surrounds a central pith). Most ferns have some type of a siphonostele. The vascular tissue is separated from the **pith** and **cortex** by an **endodermis**. The space in the vascular bundle is a **leaf gap**.
2. Make a labeled diagram of an *Adiantum* rhizome, showing the general arrangement of these tissues (Fig. 12.2). *Do not show any cellular detail*, but use outlines to indicate the following tissues: **epidermis, cortex, vascular tissue, leaf gap**, and **pith.**
3. Next, use a dissecting microscope to examine a prepared slide of a rhizome from the fern, *Polypodium*, which shows a modified form of siphonostele. A **leaf gap** is present here also, but not as obvious.
4. How do *Adiantum* and *Polypodium* differ in the appearance of their vascular tissues? (1)
5. Make a diagram of the *Polypodium* rhizome (*also without cellular detail*), showing and labeling the same tissues (Fig. 12.3).
6. Use a compound microscope to examine the vascular tissues of these two genera in more detail. Notice that the **phloem** completely surrounds the **xylem** in both. A well-developed **endodermis** separates the phloem from the cortex. This is more easily seen in *Polypodium*. Examine the small cells next to the cortex for evidence of a Casparian strip. Label the **xylem, phloem,** and **endodermis** in Figure 12.4, which is a close-up of a vascular bundle of *Polypodium*.

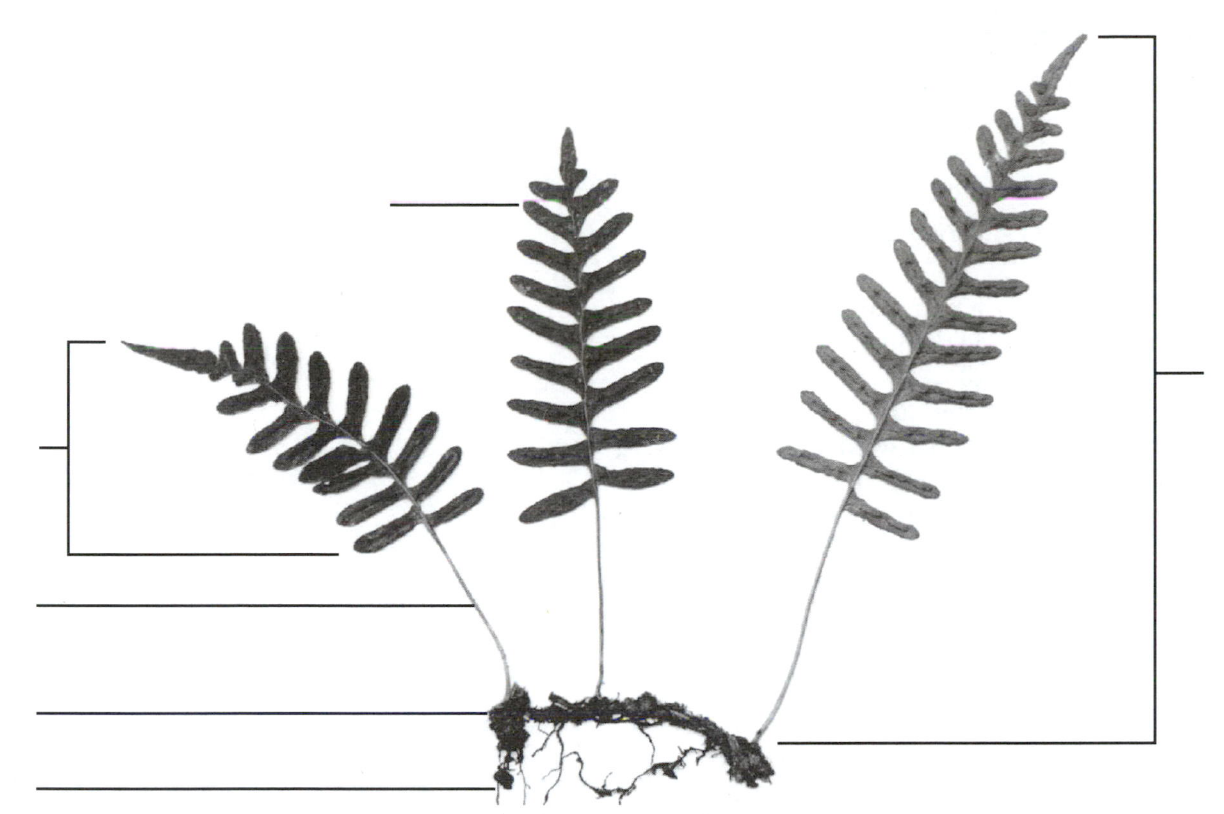

FIGURE 12.1 Herbarium mount of *Polypodium vulgare*, X 0.5.

FIGURE 12.2 *Adiantum* rhizome.

FIGURE 12.3 *Polypodium* rhizome.

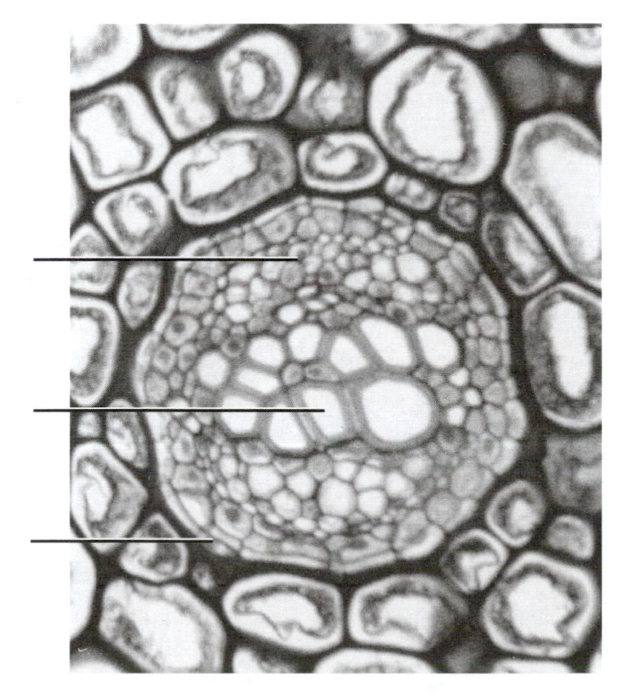

FIGURE 12.4 Vascular bundle of *Polypodium*, X 170.

Asexual Reproduction

Examine a fertile frond. On the underside of such a leaf are numerous brown spots. These are **sori** (sing. = **sorus**). Each sorus is a cluster of many **sporangia**. The sorus may be partially covered by a flap of sporophyte tissue called an **indusium**.

1. With a dissecting needle, scrape several sporangia into a drop of water on a slide, and make a wet mount. Examine these with low power. Each sporangium has a basal stalk and a terminal, more-or-less spherical, portion where the spores are produced. Find the row of thick-walled cells, the **annulus**. Notice that these cells are not thickened on the outside. As the cells of the annulus dehydrate, a negative pressure results. Eventually, this pressure is great enough to tear open the sporangium. Two special, thin-walled **lip cells** may be seen where the sporangium will tear open. After a while, the sporangium springs back to its original position, catapulting the spores into the air. This process can sometimes be seen when sporangia are placed on a clean, dry slide or when the water evaporates from a wet mount.

   Make a diagram of a single sporangium. Label the **annulus**, and if seen, **spores**, and **lip cells** (Fig. 12.5).

FIGURE 12.5 Fern sporangium.

2. Now examine a prepared slide of a fern leaf that shows sori and sporangia. Note the presence of an indusium in this species.
3. Label the **indusium, a sporangium,** and an **annulus** in the photograph of a fern sorus (Fig. 12.6).

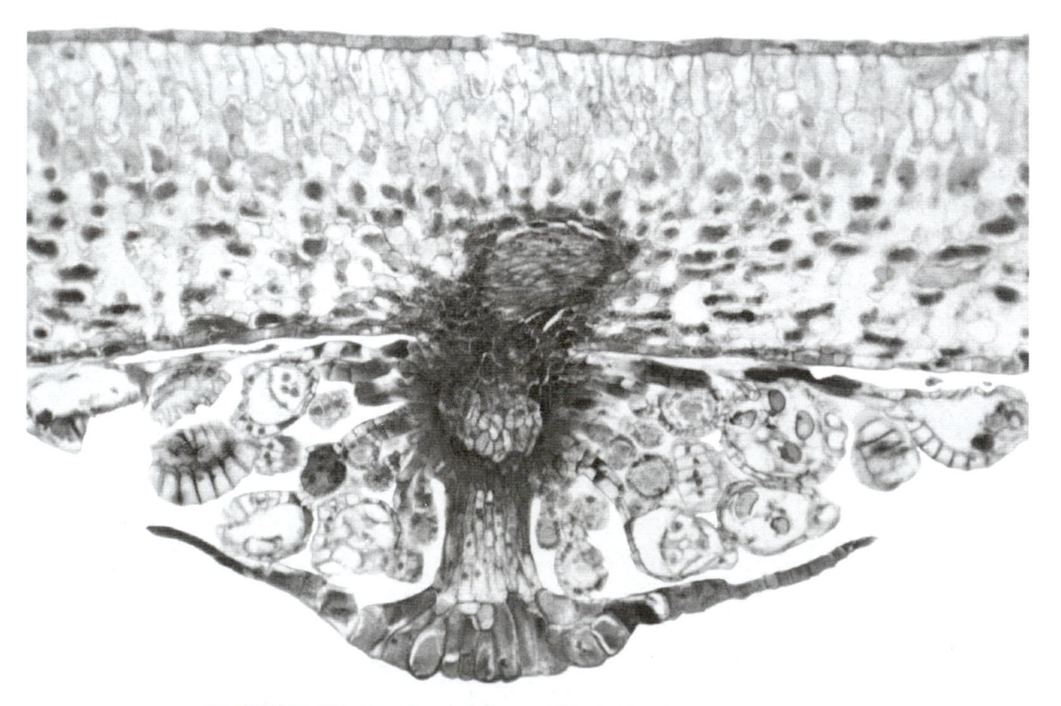

FIGURE 12.6 Fern sorus with indusium, X 80.

GAMETOPHYTE

1. If they are available, make a wet mount of a living fern gametophyte or **prothallus** (pl. = **prothalli**) using a depression slide and cover slip. Examine it with the dissecting microscope. Note the general shape, the rhizoids, and that most of it is only a single cell thick. If antheridia and archegonia are present, they will appear as dark spots.

2. Now examine a prepared slide of a prothallus. This example has both antheridia and archegonia present (some species have only antheridia *or* archegonia). If they are both present (as in our examples), the slightly elongated archegonia are usually found near the apical notch, with the spherical antheridia usually scattered on the bottom half of the prothallus. Both are found on the lower surface. Notice how they differ in shape. Label the **archegonia, antheridia, rhizoids,** and **prothallus** in Figure 12.7.

   Are these prothalli dioecious or monoecious? (2)

3. Next, examine prepared slides of longitudinal sections of fern prothalli having antheridia. The antheridia will appear as generally spherical structures. Notice the flagellated sperm cells in mature antheridia. Label an **antheridium, sperm, prothallus,** and **rhizoid** in Figure 12.8.

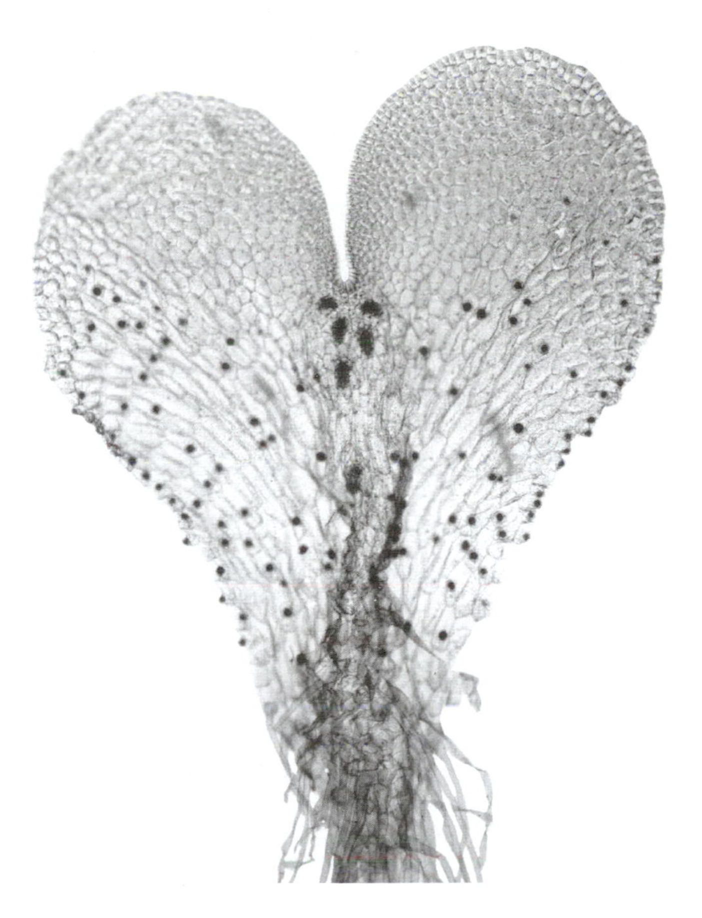

FIGURE 12.7 Fern prothallus, X 45.

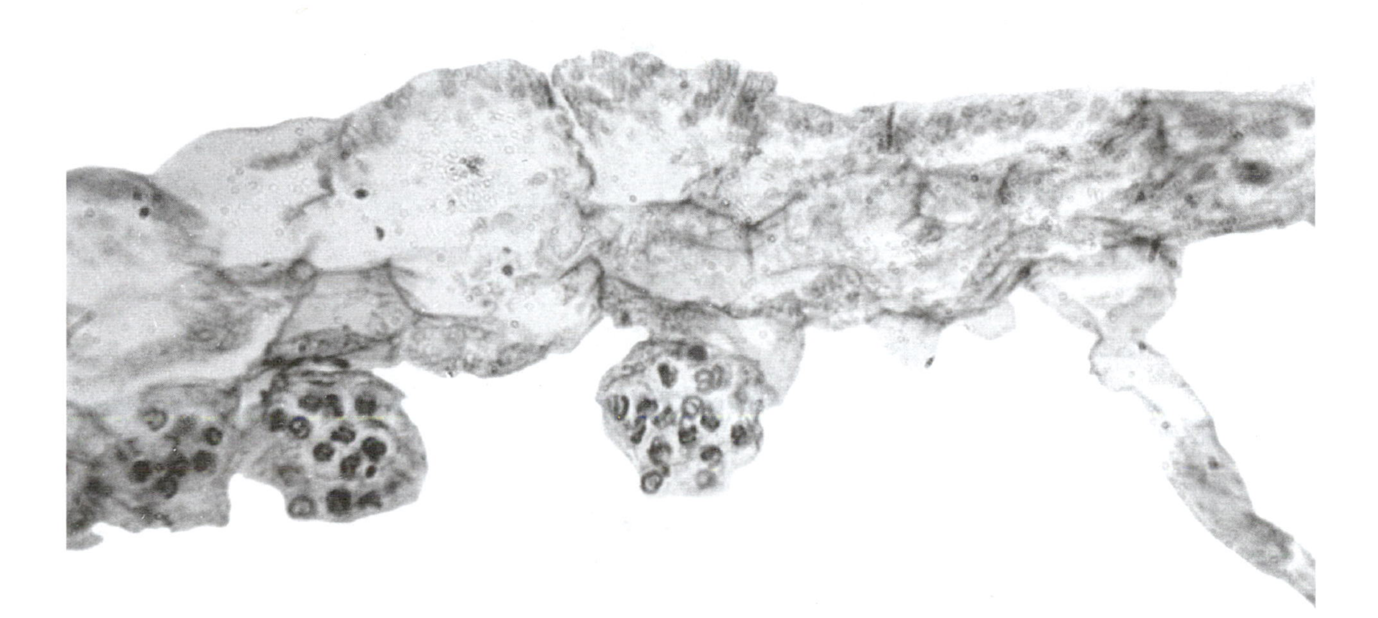

FIGURE 12.8 Longitudinal section of a prothallus with antheridia, X 400.

4. Examine a prepared slide of a fern prothallus bearing archegonia. These structures are more elongated than the antheridia. Fern archegonia have the same general structure as those of the bryophytes but are smaller. Label the **neck, venter,** and **prothallus** in Figure 12.9.

5. After the egg is fertilized, a sporophyte plant develops. Figure 12.10 shows a very young sporophyte plant.

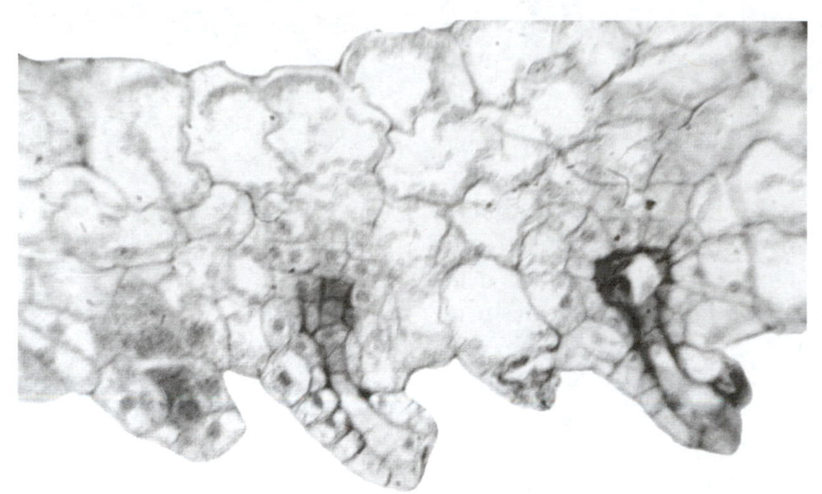

FIGURE 12.9 Longitudinal section of a prothallus with archegonia, X 260.

FIGURE 12.10 Very young fern sporophyte, X 40.

The sporophyte quickly develops a **primary leaf** and **primary root** (Fig. 12.11). Young sporophytes (such as the one shown in Fig. 12.10) receive their nourishment from the leafy gametophyte, but soon become independent, and the gametophyte dies. These primary leaves and roots are temporary and are later replaced by the more familiar fern frond and adventitious roots. Label the **prothallus, primary leaf,** and **primary root** in Fig. 12.11.

Indicate whether the prothallus, primary leaf, and primary root is haploid or diploid.

Immature fern leaves are often called *fiddleheads* or *monkey tails* (Fig. 12.12). This is in reference to the way they are coiled in the developing stage. Fern leaves do not develop in buds as do the leaves of trees and most seed plants. Observe the preserved specimen of a fern fiddlehead showing **circinate vernation** (a term used to describe this peculiar development of the young leaf).

Some of the slides of slightly older sporophytes may have small fiddleheads developing.

See if any of the living specimens have fiddleheads present.

FIGURE 12.11 Fern sporophyte with primary leaf and primary root, X 40.

FIGURE 12.12 Fern fiddlehead, X 1.

Questions

1. How does a fern prothallus differ from a moss protonema in appearance? (3)

2. How does the internal organization of the fern stem (rhizome) differ from that seen in *Psilotum*? (4)

3. Under what weather conditions are spores most likely to be released from the sporangium? What is the advantage of this? (5)

4. Bryophytes, primitive vascular plants, and ferns all demonstrated some unusual features associated with humidity and spore dispersal. What were these activities? (6)

	Dry	**Humid**
Marchantia		
Mosses		
Psilotum		
Equisetum		
Ferns		

5. Is the indusium haploid or diploid? (7)

6. The antheridia and archegonia are located on the lower surface of the prothallus. Of what benefit is this to the plant? (8)

7. Complete the following diagram of the life cycle of a fern plant. All of the various structures and events of importance are listed below. (10)

Antheridium	Mature sporophyte	Spore dispersal
Archegonium	Meiosis	Spore germination
Egg	Prothallus	Spore mother cells
Embryo	Sperm	Tetrad of spores
Fertilization	Sperm transfer	Young sporophyte
Fiddlehead	Sporangium	Zygote
	Spores	

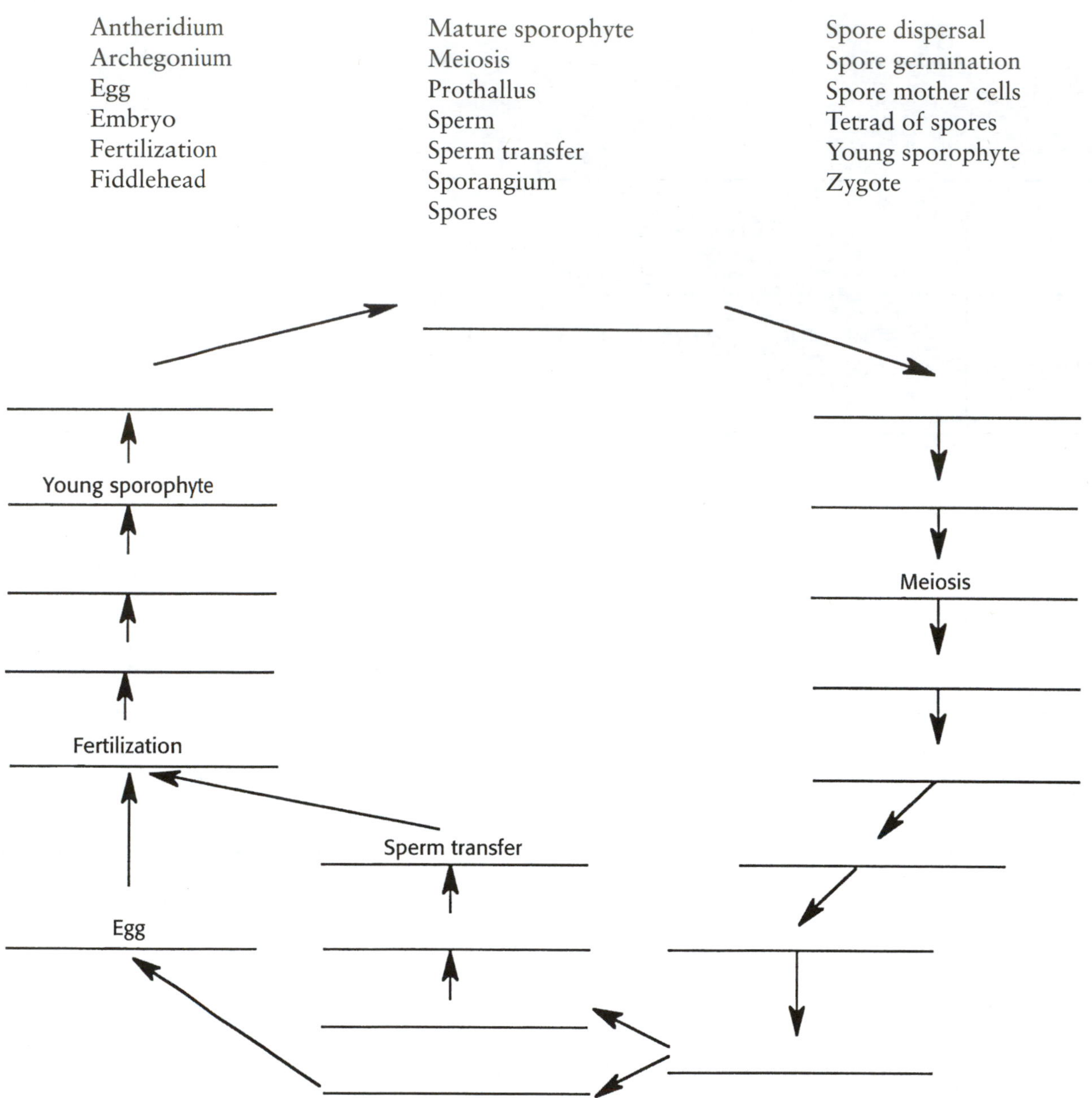

EXERCISE 13

Division Coniferophyta: Conifers

Introductory Notes

We now come to the two groups of plants that are considered to be the most advanced. These are plants that reproduce by means of seeds. Seed-bearing plants are commonly placed into two categories:

1. **Gymnosperms** produce seeds that are exposed or on the sporophyll. The word *gymnosperm* means "naked seed," in reference to this method of seed production.
2. **Angiosperms** are plants that produce their seeds inside a protective vessel, the fruit. These are the flowering plants.

There are several categories of gymnosperms. However, we will examine only one, the Coniferophyta, or cone-bearing plants, using pine as an example. Hopefully, after finishing this exercise, you will not refer to all conifers as "pine trees." This designation should be used only for those members of the genus *Pinus*. Even the designation "evergreen" is not appropriate, because some typically lose their leaves (needles) every year, and in many these needles live for only two or three years.

GOALS

After completing this exercise, the student should be able to:

- explain the main difference between angiosperms and gymnosperms.
- describe how pine leaves differ from the leaves of other conifers.
- identify the tissues of a pine leaf as seen in cross section.
- describe the structure of the following:
  1. A male pine cone
  2. A pine pollen grain
  3. A female pine cone
  4. A pine ovule
- describe the main events in the life cycle of a pine.
- demonstrate the ability to use and construct a dichotomous key.
- Identify the following genera of conifers:
  1. *Pinus*
  2. *Picea*
  3. *Juniperus*
  4. *Tsuga*
- answer the questions in the exercise.

LEAVES

Conifers are often referred to as evergreens because the leaves of most species persist for more than a single year. Fresh specimens of pine (*Pinus*) boughs are available for your examination. Look closely at these specimens, and notice the needle-shaped leaves grouped together in a basal sheath. Pines typically bear their leaves in a cluster of two to five leaves. The number of leaves in a cluster determines the shape of the leaf as seen in cross section. This number is also characteristic for the species, i.e., all individuals of any given species have the same number of leaves in a cluster.

Examine prepared slides of cross sections of the leaves of *Pinus*. Be sure to examine both the two-needle specimen and the five-needle specimen. Observe the thick-walled **epidermis** with **cuticle**. Note that at certain places the epidermis appears to be broken. Upon closer observation, you will see that rather than being broken, the epidermis is merely indented. At the base of each indentation are two **guard cells** with **stoma**. The guard cell–stomata apparatus allows the exchange of gases between the atmosphere and the leaf. The sunken stomata are an adaptation for a xerophytic type of habitat.

The vascular tissue is in a bundle in the central portion of the leaf. You should be able to identify both **xylem** and **phloem**. These are in the usual locations, with the phloem more toward the outside of the vascular bundle. The xylem and phloem are surrounded by thin-walled parenchyma cells with occasional tracheids. This tissue is known as **transfusion tissue** and apparently conducts substances between the vascular tissues and the mesophyll. The transfusion tissue is bordered on the outside by a ring of oval, thick-walled cells that comprise the **endodermis**.

Between the epidermis and the vascular cylinder, many large, oddly-shaped (often star-shaped or clover-shaped) cells can be found. These cells comprise the **mesophyll** and contain large numbers of chloroplasts. **Resin ducts** should be found in the mesophyll.

Label the **epidermis, cuticle, guard cells with stoma, xylem, phloem, transfusion tissue, endodermis, mesophyll**, and **resin duct** in *each* of the following photographs of *Pinus* leaf cross sections (Figs. 13.1 and 13.2).

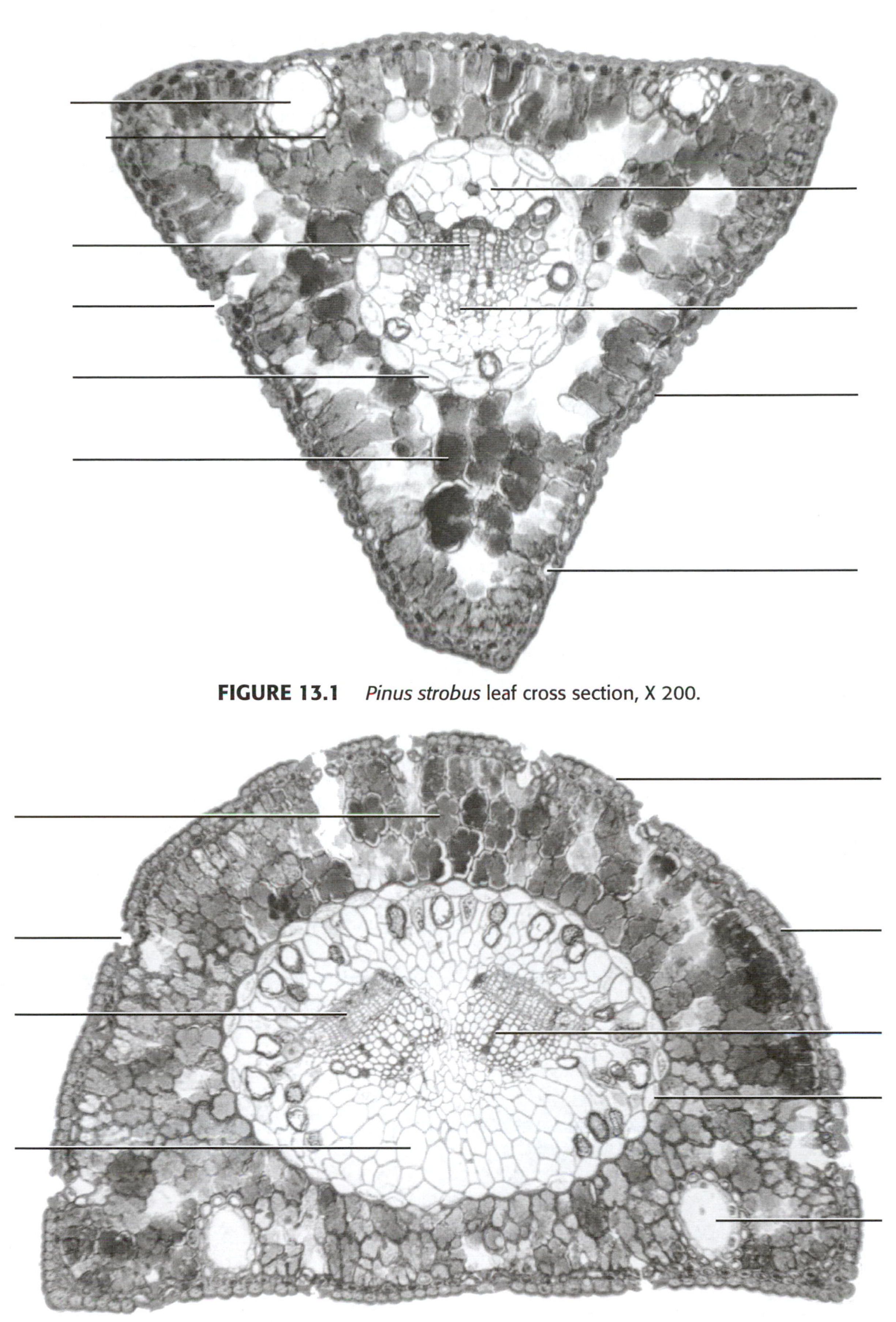

FIGURE 13.1 *Pinus strobus* leaf cross section, X 200.

FIGURE 13.2 *Pinus nigra* leaf cross section, X 130.

LIFE CYCLE

In conifers, the male and female reproductive structures are borne in separate woody strobili known as *cones*. These cones are usually produced on the same plant. Is the above situation representative of a dioecious organism or a monoecious one? (1)

Often the female cones are produced more abundantly in the upper portion of the tree.

What are two advantages of this arrangement of cones on the tree? (2)

1.

2.

Male Cones

The male cones of pine are produced in clusters near the end of a twig. Each male cone (also referred to as a pollen cone or staminate cone) consists of a central axis, and a considerable number of closely aggregated, scalelike leaves that bear **microsporangia** (Fig. 13.3).

Examine the demonstration of male pine cones. Note the relatively smooth surface texture and elongated shape. Each of the cone scales visible produces two microsporangia. What is the technical term for these cone scales? (3)

FIGURE 13.3 *Pinus strobus*, cluster of male cones, X 1.

As the male cones develop, the contents of the microsporangia differentiate into numerous microspore mother cells. These will then divide by meiosis, each giving rise to four haploid microspores. Each microspore develops, over most of its surface, a thick wall composed of an inner and outer layer. In two localized regions of the microspore wall, the outer layer separates from the inner layer, resulting in the formation of a pair of winglike structures filled with air.

The microspore germinates while still in the microsporangium. After two mitotic cell divisions, a four-celled structure is formed. This is a **pollen grain**. *The pollen grain represents an immature male gametophyte plant*. When the pollen grains are ready to be released, the male cone elongates, separating the microsporangia. The microsporangia then rupture, releasing the pollen grains which are then easily carried about by the wind. Because conifers are wind pollinated plants, massive amounts of pollen are produced each year.

Examine a prepared slide of a longitudinal section of a male pine cone. Note the arrangement of the microsporangia with the enclosed pollen grains. Examine these pollen grains with high power, and find some with well-developed bladderlike wings and the central cell with nuclei. How might these "wings" aid in pollination? (4)

Of the four nuclei that are formed during the development of the male gametophyte, two (called prothallial nuclei) deteriorate. You should be able to find pollen grains with two distinct cells of unequal size present. The larger cell is the **tube cell** with a corresponding **tube nucleus**. The smaller cell is the **generative cell** with its **generative nucleus**.

Label the **cone axis**, **microsporophyll**, **microsporangium**, and **pollen grains** in the photograph of a male pine cone (Fig. 13.4).

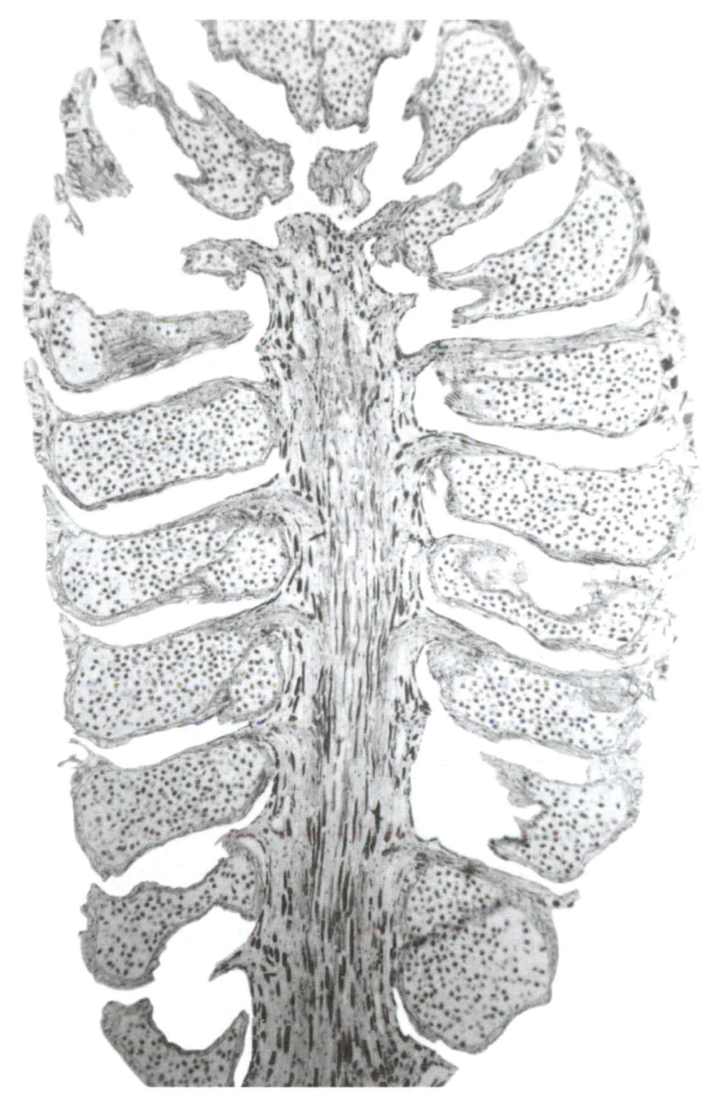

FIGURE 13.4 Longitudinal section of a male pine cone, X 20.

Make a diagram of a single pollen grain, labeling the **tube nucleus**, the **generative nucleus**, and **wings** (Fig. 13.5).

FIGURE 13.5 Pine pollen grain.

After pollination, the generative nucleus divides once by mitosis, producing two **sperm nuclei**. A tubular structure, the **pollen tube**, transports these sperm nuclei to the vicinity of the egg for fertilization. The growth of these pollen tubes apparently is stimulated by the **tube nucleus**, which leads the way and is found near the tip of the pollen tube. Figure 13.6 is a photograph of pine pollen in various stages of pollen tube development. Label a **pollen tube, tube nucleus**, and **sperm nuclei.**

Female Cones

Examine the female pine cone in its first year of development, and compare its physical features to the male cones. Note that the female cone is similar in overall construction to the male cone. However, it is evident that there are many differences. The cone scales here

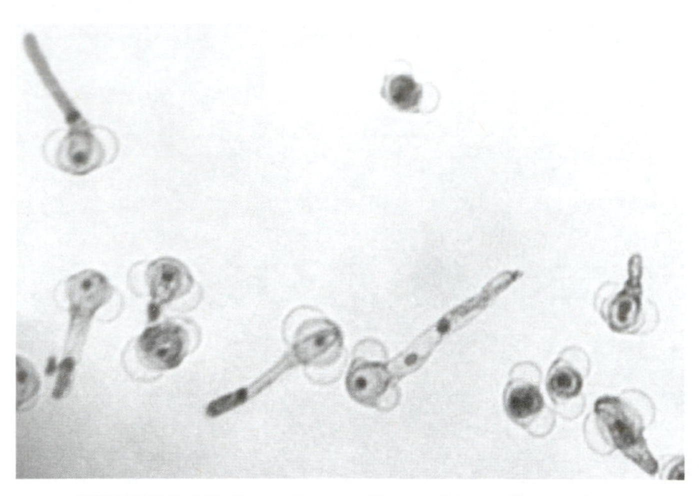

FIGURE 13.6 Pine pollen with pollen tubes in various stages of germination, X 225.

are larger than the microsporophylls. Their pointed nature may be an aid in capturing the wind-blown pollen grains. These are also more woody than the male cones. The cone scales typically bear two megasporangia, just as the microsporophylls produced two microsporangia. However, the megasporangia are not produced on megasporophylls, but on specialized structures called **ovuliferous scales.**

The megasporangium is found inside a structure called an **ovule.** In *Pinus* and other conifers, each scale produces two ovules. Each ovule produced only a single functional megaspore.

If available, examine a prepared slide of a young female pine cone with young ovules and diploid **megaspore mother cells.** This diploid cell will divide by meiosis to produce four haploid megaspores. There is a small opening in the integument, called the **micropyle,** at one end of the ovule. This is where the pollen grains sift in and come in close proximity to the female gametophyte.

Three of the megaspores will degenerate, leaving a single functional megaspore. The nucleus of this megaspore then divides many times, producing many free nuclei. Eventually, these nuclei will develop cell walls. This resulting multicellular structure is the **female gametophyte.**

Examine a prepared slide of a mature *Pinus* ovule. The ovule is surrounded by an outer integument, beneath which is the nucellus (the remains of the megasporangium).

The **female gametophyte** is in the center of the ovule. Two or more **archegonia** develop at the microphyle end of the female gametophyte. These are formed at the base of two neck cells and consist of a single layer of cells that surround a single, large egg.

Look closely for the single layer of archegonial cells surrounding the egg, which occupies the entire region inside the archegonium. An egg nucleus may not be present on the preparation.

Label the **ovuliferous scale, archegonium, egg, female gametophyte, nucellus, integument, pollen chamber,** and **micropyle** in the diagram of a *Pinus* ovule (Fig. 13.7).

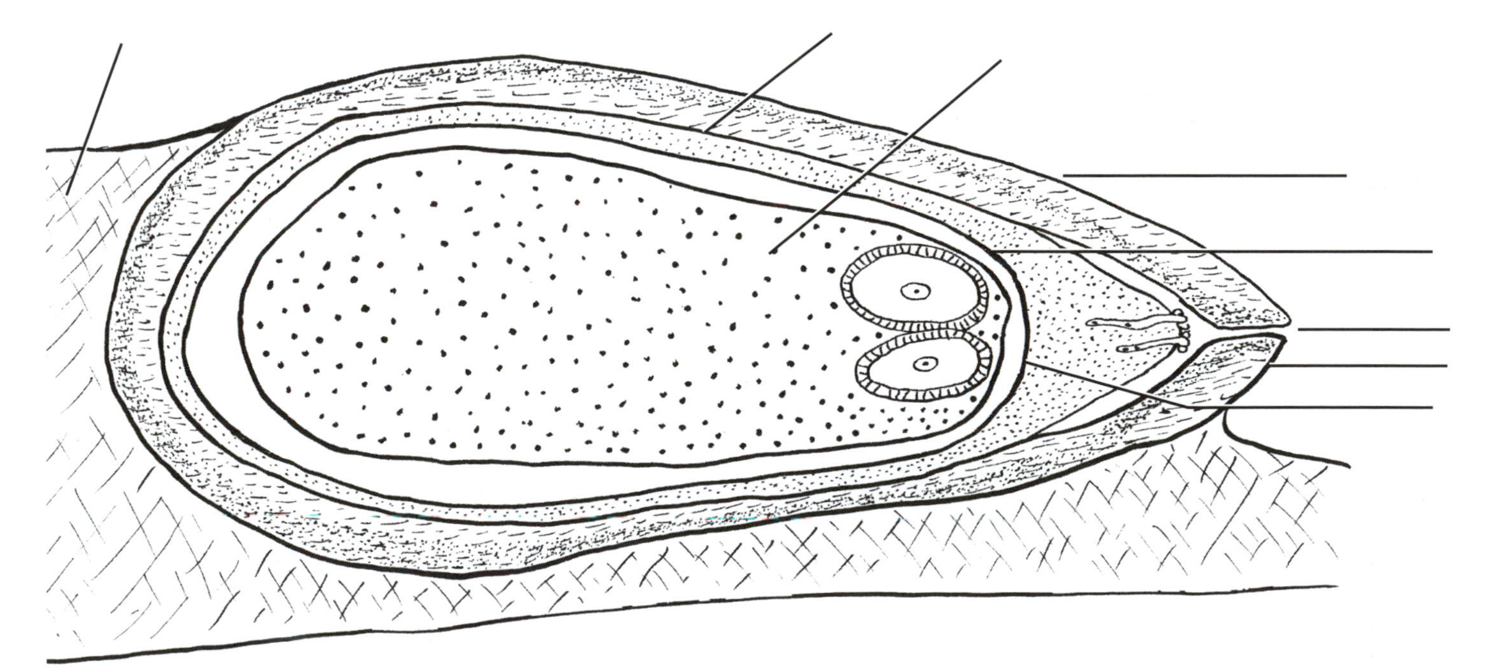

FIGURE 13.7 *Pinus* ovule, longitudinal section, X 45.

Pollination and Fertilization

At the time of pollination, the diploid megaspore mother cells have not yet divided. Development of the female gametophyte is not completed until the year after pollination. Fertilization of the egg is delayed for more than a year following pollination. In the intervening time, the pollen tubes develop and grow through the nucellus toward the archegonia. Some of our *Pinus* ovule slides show portions of these pollen tubes. After fertilization, the embryo uses the female gametophyte tissue as a source of food as it develops. Figure 13.8 shows the early stages of development of the embryo as it is pushed into the female gametophyte tissue by special cells comprising the suspensor.

Continued Seed Development and Seed Dispersal

Examine a second-year female cone, and compare its size and texture to the first-year cone (Fig. 13.9). This is about the size of the cone at the time of fertilization.

The seeds are released during the autumn of the second year. It should be mentioned that only *Pinus*, of our conifers, has this two-year development period.

Cones from several species of *Pinus* from different parts of the country are available for your examination. These are all mature, three-year female cones that have released their seeds.

Notice how the cone scales have spread, allowing seed dispersal. Examine the upper surface of the cone scale, and find the area where the seed was formed.

Pine seeds are wind disseminated and have flattened winglike extensions to aid in this dispersal. Some pine seeds have been extracted for your observation of this characteristic.

The seeds of many of our conifers (spruces, hemlocks, larches, etc.) have this dispersal method. Some of these cones are on demonstration also.

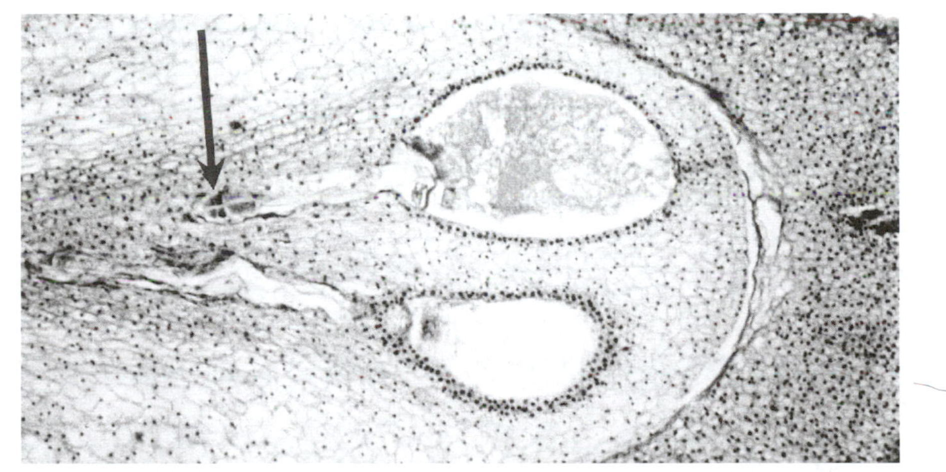

FIGURE 13.8 *Pinus* embryo (arrow) with suspensor, X 30.

FIGURE 13.9 *Pinus sylvestris*, first-year cone, second-year cone, and a mature cone that has recently released its seeds; X 2, X 0.75, X 0.5.

IDENTIFICATION OF LOCAL CONIFERS

Use the following **key to local conifers** to identify the various species of conifers present in the laboratory. A list is available for your use to check your answers. Not all of these species are native plants; some are planted as ornamentals.

1. Leaves produced in fascicles (clusters) of two or more ---------- 2
1. Leaves produced singly (they may have alternate, opposite or whorled arrangement) ---------- 8

    2. Leaves produced in fascicles of two to five; usually with a thin sheath at the base ---------- genus *Pinus* ---------- (pines) ---------- 3
    2. Fascicles containing more than five leaves, produced at the end of short, stout spur shoots; deciduous ---------- genus *Larix* ---------- (larches)
    Larix laricina, the Eastern Larch or Tamarack, is the native species of the region, but the European Larch (*L. decidua*) is more common.

3. Leaves slender, five per fascicle; cones slender and flexible -- *Pinus strobus* -------- (Eastern White Pine)
3. Leaves two per fascicle; cones rigid ---------- 4

    4. Leaves generally less than 3 inches long ---------- 5
    4. Leaves more than 4 inches long ---------- 7

5. Leaves twisted; plant treelike ---------- 6
5. Leaves not twisted; plant shrubby ---------- *Pinus mugo* ---------- (Swiss Mountain Pine)

    6. Leaves very short (1/2 to 1½ inches); cones curved ----- *Pinus banksiana* ------ (Jack Pine)
    6. Leaves generally longer than 1½ inches, with a bluish cast; cones with a flat exposed scale surface ---------- *Pinus sylvestris* ------ (Scotch Pine)

7. Leaves dull, thick, not brittle (they do not break cleanly when bent double; leaf sheath shorter; cones with a small prickle at the tip of each scale--- *Pinus nigra*---------- (Austrian Pine) ---- {**A**}
7. Leaves more shiny, slender, brittle (they break easily when bent double); leaf sheath longer, cones smooth ---------- *Pinus resinosa* ------- (Red Pine) ------- {**A**}

    {**A**} It is often difficult to differentiate between these two species. If possible, examine the leaves of both to make a comparison. The small prickle at the tip of each cone scale of the Austrian Pine is an excellent identifying feature. The bud scales of the Red Pine are more of an orange-brown color as compared to the more silvery color of those of Austrian Pines.

    8. Leaves opposite or whorled ---------- 9
    8. Leaves one per node, usually arranged in a spiral pattern ---------- 11

9. Leaves all scalelike (short, flat, closely appressed), not easily observed ---------- *Thuja occidentalis* --- Eastern Arborvitae or Eastern White Cedar
9. At least some of the leaves awl-like (sharply pointed)------ *Juniperus* ---------- (junipers) -------- 10

    10. Branchlets bearing only awl-shaped leaves which are in whorls of three ---------- *Juniperus communis* ------- (Common Juniper)
    10. Branchlets with both awl-shaped and scalelike leaves ---------- *Juniperus virginiana* ------- (Eastern Red Cedar) {**B**}

    {**B**} The Pfitzer Juniper (*J. chinensis*) has similar characteristics but is a low-growing shrub.

11. Leaves more or less four-sided in cross section, often stiff and set close together on the twig; older branches show well-developed stalks to which the leaves were once attached ---------- genus *Picea* ---------- (spruces) ---------- 12
11. Leaves flattened in cross section, or with other combinations of characteristics ---------- 14

12. Leaves very stiff and pointed, radiating at nearly right angles to the twig, releasing a strong odor when crushed; cones 2½ to 4 inches long - - - - - - *Picea pungens*- - - - - - - (Colorado Blue Spruce)
12. Leaves more flexible; cones not as above - 13

13. Leaves shiny, dark green; branches drooping conspicuously; cones 4 to 7 inches long - *Picea abies* - - - - - - - - - - (Norway Spruce)
13. Leaves generally dull, not dark green, crowded onto the upper side of the twig; cones about 2 inches long - *Picea glauca* - - - - - - - - - (White Spruce)

14. Leaves abruptly narrowed onto short stalk - 15
14. Leaves not abruptly narrowed onto short stalks - 16

15. Leaves blunt at the tip, about 1/2 inch long, two conspicuous white lines on the under surface; cones about 3/4 inch; plant treelike - - - - - - - - - - - - - - - - - *Tsuga canadensis* - - - - - (Eastern Hemlock)
15. Leaves pointed at the tip, white lines lacking; plant shrubby - genus *Taxus* - - - - - - - - - (yews) - - - - - - - - - - {C}

{C} The native yew is *Taxus canadensis* and is a small, low-growing shrub. A number of cultivated species, such as the Japanese Yew (*T. cuspidata*) and the English Yew (*T. baccata*) are commonly planted as ornamentals. Yews differ from other conifers in that they are dioecious (there are male plants and female plants). The seeds are not produced in cones either. Rather, the female plant bears pink, fleshy, cuplike structures that partially surround a single seed. This fleshy structure (called an aril) is edible when ripe, but the seeds are very toxic. Because of this, female plants are not planted as often as the male plants.

16. Leaf scars more elliptical and slightly raised from twig surface; cones with a conspicuous three-pronged bract, which extends beyond the tip of the cone scale - - *Pseudotsuga menziesii* (Douglas Fir)
16. Leaf scars (on young twigs) nearly circular, not raised (older twigs have a smooth surface); cones do not remain intact and are difficult to obtain - - genus *Abies* - - - - - - - - - (firs) - - - - - - - - - - - 17

17. Leaves 2 to 3 inches long, bluish green, mostly curving upward, lemon-scented - *Abies concolor* - - - - - - - (White Fir)
17. Leaves generally not more than 1 inch long, green, no citrus aroma - *Abies balsamea*- - - - - - - (Balsam Fir)

Identification of laboratory specimens:

1. ______________________ 9. ______________________

2. ______________________ 10. ______________________

3. ______________________ 11. ______________________

4. ______________________ 12. ______________________

5. ______________________ 13. ______________________

6. ______________________ 14. ______________________

7. ______________________ 15. ______________________

8. ______________________ 16. ______________________

Questions

1. How do the leaves of *Pinus* differ from those of other conifers? (5)

2. Conifers are gymnosperms. What does this mean? (6)

3. What is the function of the mesophyll? (7)

4. Pine leaves are adapted for conserving water. What are three adaptations that can be seen, either by examining the entire leaf or by looking at a cross section? (8)

   a.

   b.

   c.

5. In a botanical sense, what is a pollen grain? (9)

6. What is an ovule? (10)

7. The parts of an ovule identified include the following structures. Indicate whether each is haploid or diploid. (11)

   a. Archegonium ____________________

   b. Egg ____________________

   c. Female gametophyte ____________________

   d. Integument ____________________

   e. Nucellus ____________________

   f. Ovuliferous scale ____________________

8. What are two features that make it easy to identify the following:

   Spruce (*Picea*)? (12)

   a.

   b.

   Hemlock (*Tsuga*)

   a.

   b.

   Juniper (*Juniperus*)

   a.

   b.

9. The development of a seed represents another major step in the evolution of plants. What are some advantages of having seeds? (13)

EXERCISE

14

Division Anthophyta: Flowering Plants — The Flower

Introductory Notes

Flowering plants, often called angiosperms, represent the most advanced group of plants. These are not only the most advanced plants, but they are also the most abundant and familiar of the plant groups, with at least 250,000 species. Angiosperms differ from the conifers and other gymnosperms by having their ovules and seeds enclosed within a protective structure, the pistil, which later develops into the fruit. The reproductive structure is no longer a strobilus or cone, but a flower.

FLORAL PARTS (EXTERNAL MORPHOLOGY)

A flower may be viewed as a specialized twig or stalk with highly modified leaves at its tip. The basic features of a complete flower include four types of structures, each attached at higher levels of the floral axis or **receptacle.**

The outermost and lowest parts, enclosing the other floral parts during the bud stage, are the **sepals** (collectively referred to as the **calyx**). These are usually green and resemble small leaves.

The appendages attached just above the sepals are the **petals** (collectively known as the **corolla**). Petals are usually colorful or white leaflike structures that usually extend beyond the other floral parts. The bright, contrasting color of the petals attracts animals to the flower that aid in cross pollination. The calyx and corolla together comprise the **perianth** of the flower.

The third group of appendages are the **stamens** (collectively called the **androecium**), are attached just above the base of the petals. Each stamen, a male reproductive structure, typically consists of two parts: (1) an elongated, basal stalk, the **filament** and (2) a somewhat lobed, expanded structure at the free end of the filament, the **anther**. The anthers are where the pollen is produced.

Attached at the uppermost site and in the center of the flower is (are) the female reproductive organ(s), the **pistil**(s). Each pistil is composed of one or more modified leaves called **carpels.** A pistil typically is composed of three parts: (1) a swollen, basal region or **ovary**, (2) a terminal **stigma**, and (3) the **style** that connects the first two parts. The stigma is the part of the pistil that receives pollen during pollination. The ovary is where the **ovules**, which will develop into seeds, are produced. After pollination, the ovary will develop into the fruit that encloses the seeds.

A pistil may be either **simple** or **compound.** A simple pistil is composed of a single carpel, while compound pistils are composed of two or more carpels fused together. The female reproductive structures are collectively referred to as the **gynoecium.**

The sepals and petals are known as **accessory** floral parts, since these are not absolutely necessary for reproduction. The anthers and pistils are called the **essential** parts. A **complete** flower is one that has all four floral parts. An **incomplete** flower is one that lacks any one of these. **Perfect** flowers are those with both male and female reproductive structures, while **imperfect** flowers are unisexual.

GOALS

After completing this exercise, the student should be able to:

- explain the structure of a typical complete flower.
- identify, from either a photograph or a diagram, the various floral structures studied.
- differentiate between primitive and advanced floral structures, when presented with a list of such features.
- distinguish between a dicot flower and a monocot flower, and explain these differences.
- answer the questions in the exercise.

FLORAL EVOLUTION

Even though the flowering plants are the most recent group of plants to develop, certain evolutionary trends can be seen within the group. The classification of flowering plants into the various orders and families is based primarily on floral structure. The flowers of these orders and families show certain evolutionary trends. Most classification schemes categorize these plants from the more primitive members to the more complex. The following list gives some advanced and primitive features for certain characteristics.

PRIMITIVE	ADVANCED
Receptacle elongated or convex	Receptacle flattened or concave
Floral parts spirally arranged on the receptacle	Floral parts in whorls on the receptacle
Numerous floral parts	Few floral parts (less than ten)
Floral parts not fused together	Floral parts fused
Flowers radially symmetrical	Flowers bilaterally symmetrical
Flowers hypogynous*	Flowers epigynous*

* Hypogynous flowers have the sepals, petals, and stamens attached below the ovary, while epigynous flowers show the reverse arrangement. An intermediate condition also exists which is referred to as perigynous.

Compare these points as you examine each flower. A tally sheet is included at the end of the exercise, with each condition given a numerical value. Analyze each flower, rating each condition, and obtain a sum for each specimen. When you are through, compare the totals. Those flowers with the highest totals should have the most primitive features. A water lily flower, which is on demonstration, is used as an example of a flower with many primitive features (Fig. 14.1).

Previous exercises covering plant anatomy emphasized differences between monocots and dicots. These two groups differ in their general flower structure also. Typically, monocots have their floral parts in some multiple of three (usually three or six). Dicots usually have their flower parts in multiples of either four or five.

FLOWER STRUCTURE

Tulip (*Tulipa* sp.) or Gladiolus (*Gladiolus* sp.)

Notice that the petals and sepals are both brightly colored. The outer whorl is the calyx; the inner whorl is the corolla.

FIGURE 14.1 *Nymphaea odorata*, water lily, X 0.5.

How many sepals are present?

How many petals? (1)

What type of symmetry does the flower have? (2)

Note that in the tulip, the sepals, petals, and stamens are all attached below the ovary, making the tulip an hypogynous flower. The gladiolus is epigynous, however, and the floral parts are attached above the ovary.

How many stamens comprise the androecium? (3)

Remove an *intact* stamen and make a diagram of it, labeling the **anther** and the **filament** (Fig. 14.2).

FIGURE 14.2 *Tulipa* or *Gladiolus* stamen.

In the center of the flower is a conspicuous compound pistil. How many lobes does the stigma have? (4)

The number of stigmatic lobes usually indicates how many carpels make up the pistil. Using a razor blade, cut a thin cross section of the ovary. Examine this with the dissecting microscope. These plants have compound pistils (more than one carpel fused together). The numerous, generally colorless structures are **ovules**. The ovules develop in cavities called **locules**.

How many locules are present? (5)

The structure to which these ovules are connected is called the **placenta**.

Make two labeled diagrams: (1) an intact pistil, labeling the **stigma**, **style**, and **ovary** and (2) an ovary cross section, labeling the **placenta**, **ovule**, **locule**, and **carpel** (Fig. 14.3).

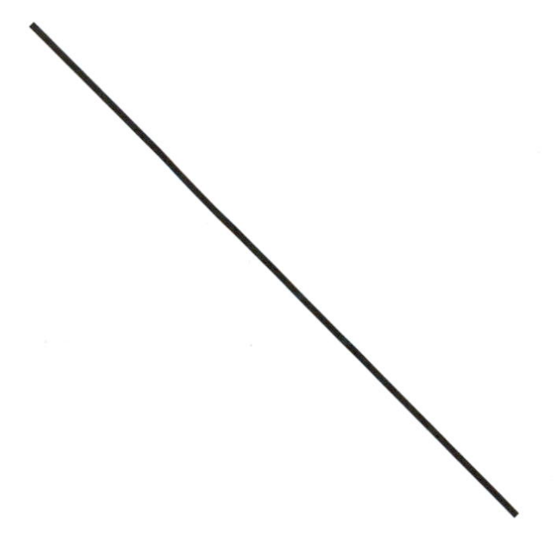

FIGURE 14.3 *Tulipa* or *Gladiolus*, intact pistil and ovary cross section.

Are these examples of dicot or of monocot plants? (6)

Florist's or cultivated geranium (*Pelargonium* sp.)

Notice the five green, leaflike sepals and the five (usually) colored petals. Are these structures free or fused? (7)

1. Remove the perianth and find the stamens. How are they arranged? (Some of the stamens may be lacking anthers — they are easily broken off.) (8)

Wild geraniums commonly have ten stamens in two sets of unequal lengths. How many stamens are present in *Pelargonium*? (9)

2. Remove the stamens to find the pistil. Is *Pelargonium* hypogynous or epigynous? (10)

3. Make a labeled diagram of an *intact* pistil of *Pelargonium* (Fig. 14.4).

4. Cut it off at its base. Is it a simple or compound pistil? (11)

5. What evidence did you use to answer the previous question? (12)

6. Make a cross section of the ovary. Notice that it is composed of five carpels. Try to locate the enclosed ovules.

FIGURE 14.4 *Pelargonium* pistil.

Rose (*Rosa* sp.)

Examine the entire flower, counting the parts of the calyx and corolla, and noting if these parts are fused or not. (13)

The rose will be used to demonstrate the theory that stamens are modified leaves (an alternative point of view states that the petals are stamens that have lost their anthers). Carefully remove the petals one by one, starting with the outermost ones, and place them in a row. Notice that the petals get progressively smaller. Eventually you will find small petal-like structures with poorly developed anthers at the top. As you continue, the anthers will become better developed, and the petal appearance will disappear.

Diagram three stages of this sequence, include a well-developed petal, an intermediate stage, and a typical stamen (Fig. 14.5). *Identify each stage.* On the intermediate stage, label the **rudimentary anther** and the **petal.** Label the **anther** and **filament** of the typical stamen.

FIGURE 14.5 *Rosa*, showing the transition from leafy structure to stamen.

Make a longitudinal section through the remaining receptacle region of the flower, and examine the cut surface. You should see a green structure formed by the fusion of the bases of the sepal, petals, and stamens. This cup-shaped structure is called an **hypanthium.** The sepals, petals, and stamens all arise from the rim of the hypanthium — an example of a **perigynous** flower. The numerous, simple pistils are enclosed within the hypanthium cup.

Make a diagram of this portion of the flower, labeling the **hypanthium, stamens,** and **pistils** (Fig. 14.6).

FIGURE 14.6 *Rosa*, longitudinal section through the hypanthium.

Snapdragon (*Antirrhinum majus*)

1. Examine a snapdragon flower.
   a. What kind of symmetry does the flower have? (14)

   b. How many sepals comprise the calyx? (15)

2. Notice that the petals are fused. The corolla tube is divided into an upper and a lower lobe. How many petals are in each part? (16)

This flower is very highly modified for insect pollination. It takes a strong insect, such as a bumblebee, to force its way into the opening between the petals. If you open the corolla, you will see the anthers and perhaps the tip of the pistil.

3. Pull off the entire corolla tube and slit it open longitudinally. Note that the stamens are attached to the petals.
   a. How many stamens are present? (17)

   b. Is the flower hypogynous or epigynous? (18)

   c. Is the pistil simple or compound? (19)

4. Using a razor blade make a cross section of the ovary, and examine it with the dissecting microscope. How many carpels are present? (20)

5. Make a labeled diagram of a cross section of the snapdragon ovary, labeling the **ovary wall, placenta, ovules,** and **locules** (Fig. 14.7).

FIGURE 14.7 *Antirrhinum*, cross section of the ovary.

Daisy, Chrysanthemum (*Chrysanthemum* sp.)

1. Count the number of sepals in the calyx. Do the same for the petals in the corolla.

   Unless you obtain the numbers 0 and 5, you have made a very common mistake. The entire structure you are examining is not a single flower, but is a group of flowers (**inflorescence**). This type of inflorescence is called a head and has a very short, compressed central axis. (Compare this to the snapdragon stalk, another inflorescence, but much more elongated.) Daisies, sunflowers, asters, black-eyed-Susans, dandelions, and many others belong to a family of flowering plants that are commonly called composites.

   The daisy type of composite head has two types of flowers, outer **ray flowers** and central **disk flowers**. The green structures at the base of the inflorescence are bracts, known as an **involucre**.

2. Cut the flower head in half longitudinally, and examine the cut surface. The small oval structures at the base of each individual flower is the ovary. Remove an intact ray flower, being sure to get the basal ovary. Examine the flower closely with the dissecting microscope. Notice the absence of any sepals and the fused petals.

   a. What type of symmetry does the ray flower show? (21)

   b. Is the flower hypogynous or epigynous? (22)

   There are no stamens, but there is a single compound pistil. However, in these daisy-type plants, these ray flowers are usually sterile. Find the other parts of the pistil.

3. Make a diagram of a single ray flower, labeling the **corolla, ovary, style** and **stigma** (Fig. 14.8).

FIGURE 14.8 *Chrysanthemum*, ray flower.

4. Examine the region of the disk flowers, and note that they are in several stages of development. Where are the *least* mature ones located? (23)

5. Remove a mature, intact, disk flower, and examine it with the dissecting microscope. Note that the petals are all fused together to form a tube.

   a. How many petals are fused together to form the corolla? (24)

   b. What type of symmetry does this flower show? (25)

6. Carefully split the corolla tube lengthwise to find the stamens. There are five of them, and the anthers are fused together in a ring surrounding the style. The short filaments are attached to the petals. There is a single, compound pistil here also. What evidence is there that the pistil is compound? (26)

Dandelion (*Taraxicum officinale*)

The dandelion is also in the composite family and has a flower structure that is basically the same as the daisy.

1. Examine the dandelion head, and notice that there is only one type of flower present. Remove some intact flowers, and observe them with the dissecting microscope. Do they resemble the ray flowers or the disk flowers of the daisy? Why? (27)

   Each flower has five petals, the points of which are easily seen at the tip of the corolla. Both stamens and a pistil are present and are arranged in a pattern similar to that seen in *Chrysanthemum*. The ring of bristles seen at the top of the ovary is called a **pappus**. This will develop into the familiar fluffy tuft which is seen when the plant produces seeds.

2. Make a labeled diagram of a single dandelion flower, showing and labeling the **corolla, pappus, ovary, stigma, style**, and **anthers** (Fig. 14.9).

FIGURE 14.9 *Taraxicum*, single flower.

DEVELOPMENT OF THE MALE AND FEMALE GAMETOPHYTES

Obtain a prepared slide of a cross section of a tulip or lily bud cross section. Examine this with the dissecting microscope to locate the four main floral parts. Then use the compound microscope to examine the anther cross sections.

Pollen Development

A young anther contains four microsporangia, two on each side of a connective (an extension of the filament). At this early stage, each microsporangium contains many microspore mother cells. Each microspore mother cell then divides by meiosis, producing four microspores that are haploid in their chromosome content.

Each microspore develops a cell wall and enlarges. The nucleus divides once by mitosis, producing a two-celled pollen grain. During this time of development, the cells separating each pair of microspores break down, and two pollen sacs, one on each side of the connective, are formed.

When the pollen grains are fully developed, some specialized cells along the side of the anther (**dehiscent cells**) dry out, and the pollen can then escape. Figures 14.11–14.14 show different stages of this process. Note the two-celled pollen grains in the last figure.

Locate some pollen grains that show this two-celled condition. As with the pine, the larger cell is the **tube cell**, and the smaller one is the **generative cell** (with similarly named nuclei). As in the conifers, these pollen grains represent the male gametophyte stage of the life cycle.

Make a diagram of one of these pollen grains, labeling the **tube cell, tube nucleus, generative cell** and **generative nucleus** (Fig. 14.10).

FIGURE 14.10 Mature pollen grain.

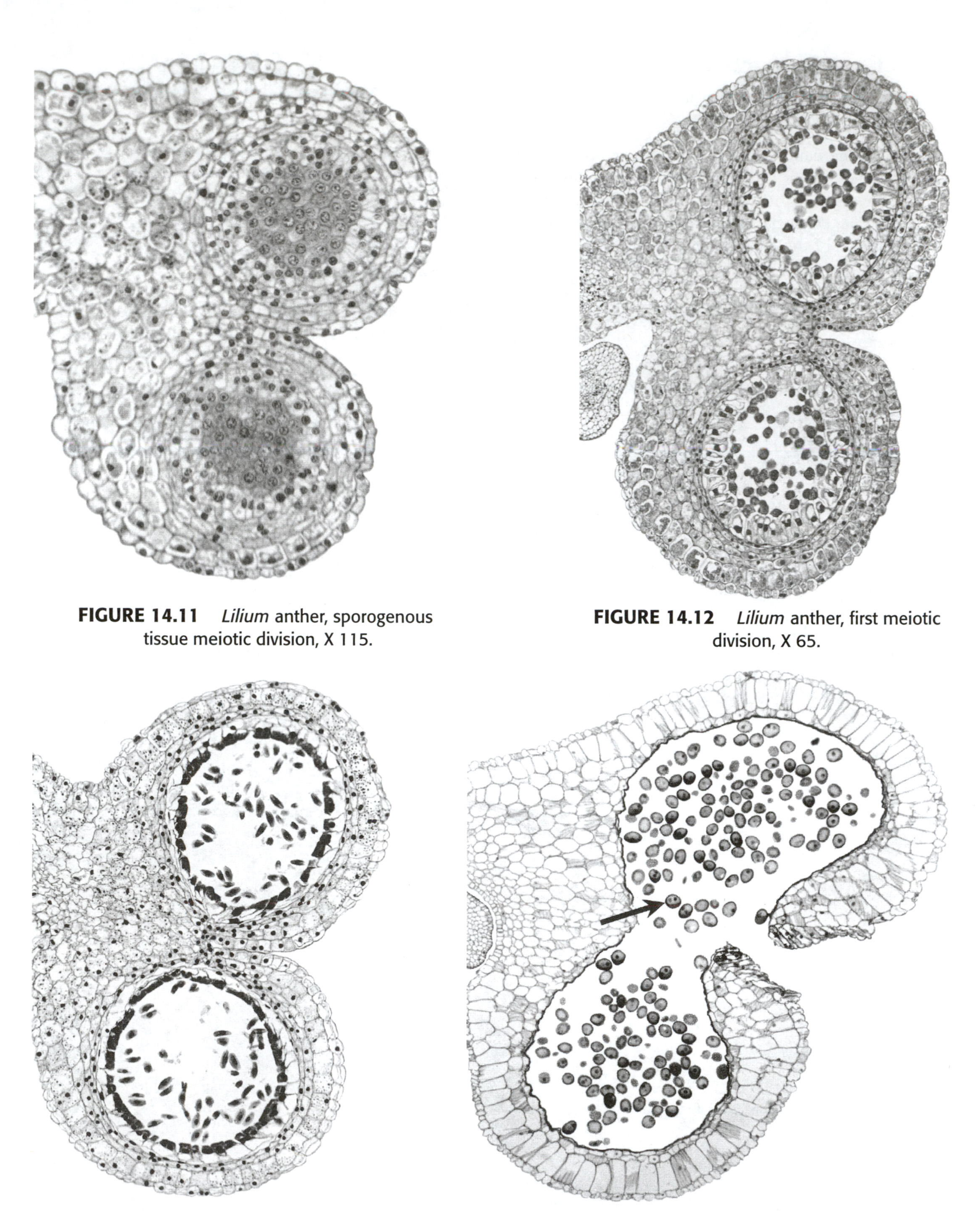

FIGURE 14.11 *Lilium* anther, sporogenous tissue meiotic division, X 115.

FIGURE 14.12 *Lilium* anther, first meiotic division, X 65.

FIGURE 14.13 *Lilium* anther, microspores, X 65.

FIGURE 14.14 *Lilium* anther, pollen grains, X 55.

Ovule Development

Very early in its development, an ovule consists almost entirely of a megasporangium or nucellus. These may appear only as small bumps on the placenta. A single, large, diploid, megaspore mother cell develops in each megasporangium (in contrast to the large number of microspore mother cells found in a microsporangium). This megaspore mother cell divides by meiosis, producing four haploid megaspores. Just as in the coniferous ovule, three of these deteriorate. The surviving megaspore then develops into the female gametophyte, which in the flowering plants is known as the **embryo sac**.

As the development continues, a layer or layers of cells that surround the embryo sac also develop. These are the **integuments**. The integuments do not completely surround the embryo sac, but leave a small pore or opening at one end. This opening is known as the **micropyle**.

The female gametophyte typically forms after three mitotic divisions, and contains eight nuclei, four at each end. The maturing process involves the migration of one nucleus from each group of four to the central region of the embryo sac. These two nuclei are the **polar nuclei**. This leaves three nuclei at the end near the micropyle. These three nuclei comprise the **egg appartatus**. One of them is the egg, the other two are called **synergids**. At the opposite end of the embryo sac are three more nuclei, the **antipodals**. Cell membranes and then thin cell walls usually develop around the antipodals and the three cells of the egg apparatus. This also happens to the polar nuclei, forming a binucleate **central cell**.

Label the **antipodals**, **polar nuclei**, **egg**, and **synergids**, in the mature embryo sac, diagrammed in Figure 14.15.

The majority of flowering plants have the type of ovule development just described. The lily, which you will look at, is different. It still forms eight nuclei, which have the same names (egg, synergids, antipodals, and polar nuclei), but the process and chromosome content of the cells is different. The lily is usually used for this observation because the ovules are large and easily seen.

Examine prepared slides of mature lily (*Lilium*) ovules within the ovary, and try to locate these structures. Because these sections are of fairly large objects,

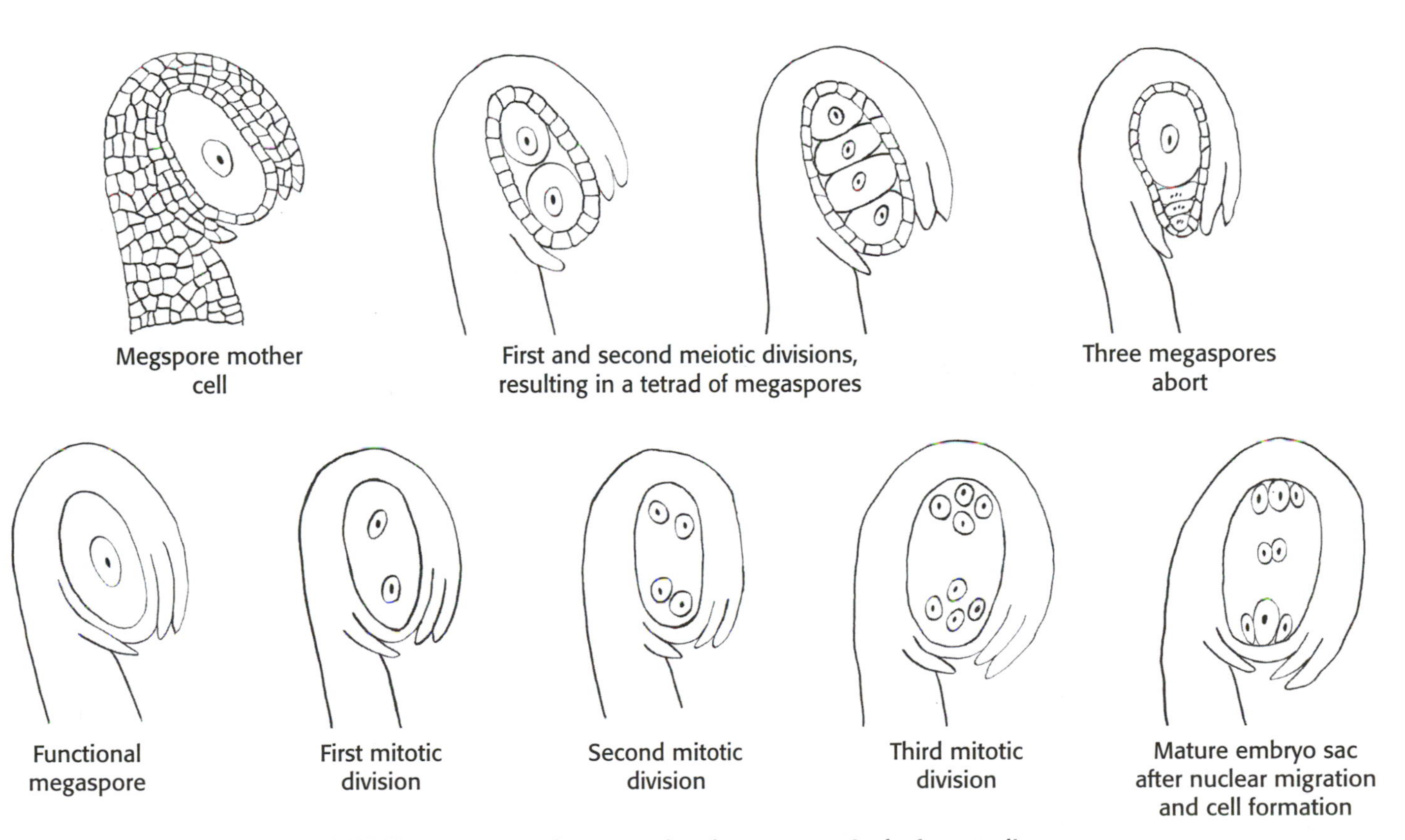

FIGURE 14.15 Embryo sac development typical of most dicots.

it will be necessary to examine more than one section or slide, and try to make a composite picture in your mind about the overall structure. Figure 14.16 is a diagramatic representation of the events associated with the formation of the embryo sac in the lily.

Notice that all four megaspores remain viable (in most other plants, three of these abort). Three of the megaspores then migrate to the end of the embryo sac that is a way from the micropyle. These three nuclei fuse, forming a single triploid (3N) nucleus. After this, the embryo sac has two nuclei; one is triploid and the other haploid. Both nuclei then divide twice by mitosis. This results in four triploid nuclei at the end away from the micropyle and four haploid nuclei at the micropylar end. One of each of these then migrates to the center of the embryo sac. These are the polar nuclei. The embryo sac now has three triploid antipodals, one triploid polar nucleus, on haploid polar nucleus, two haploid synergids, and a single haploid egg. Again, after migration, a membrane and a thin cell wall forms around the various nuclei.

The polar nuclei may fuse prior to fertilization, resulting in a 4N central cell. Double fertilization then results in a diploid zygote and a pentaploid (5N) endosperm nucleus. Compare the diagrams below with Figure 14.15. Figures 14.17–14.19 show some stages of Lilium embryo sac development.

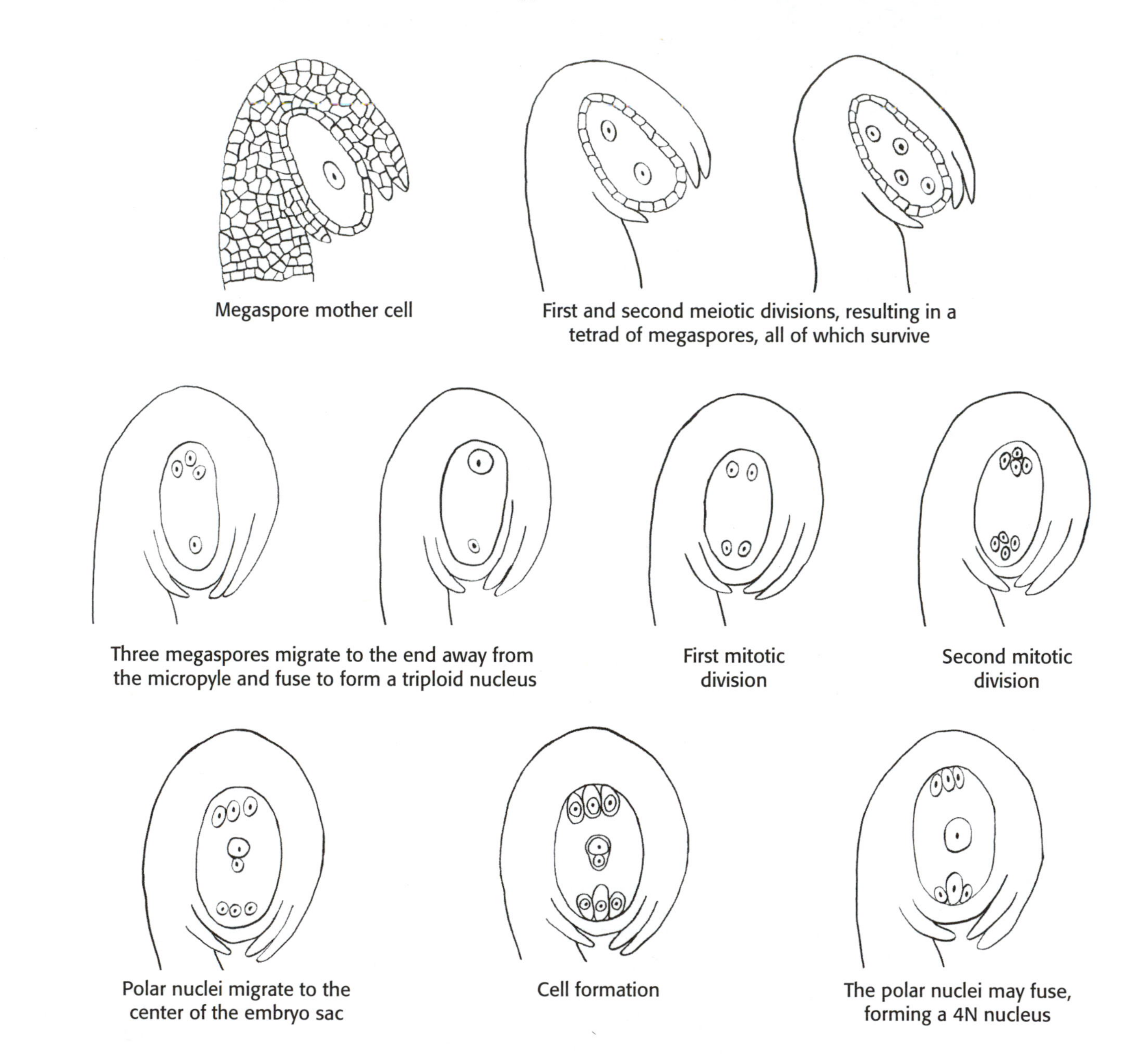

FIGURE 14.16 *Lilium*, embryo sac development.

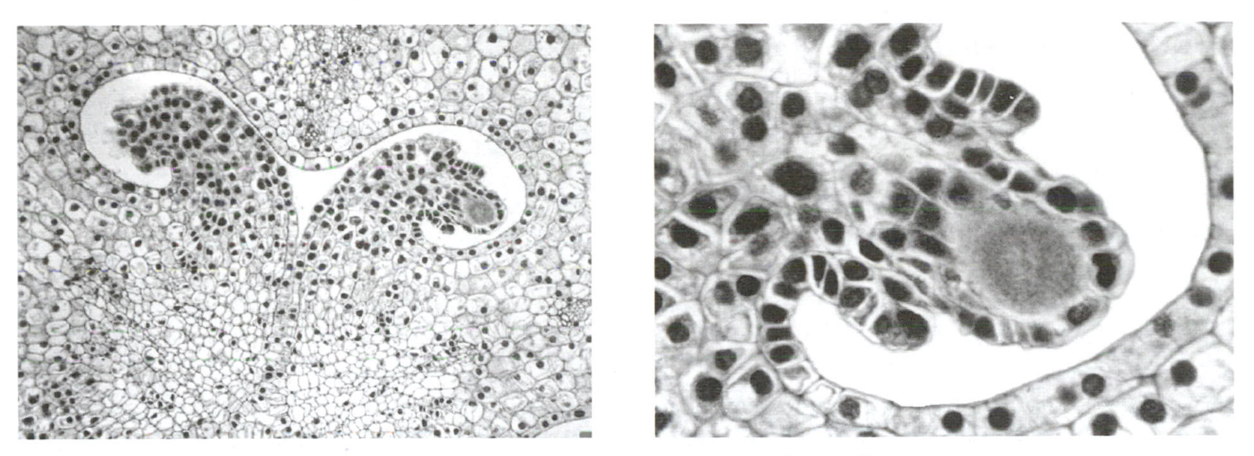

FIGURE 14.17 *Lilium* ovule, megaspore mother cell, X 40, X 160.

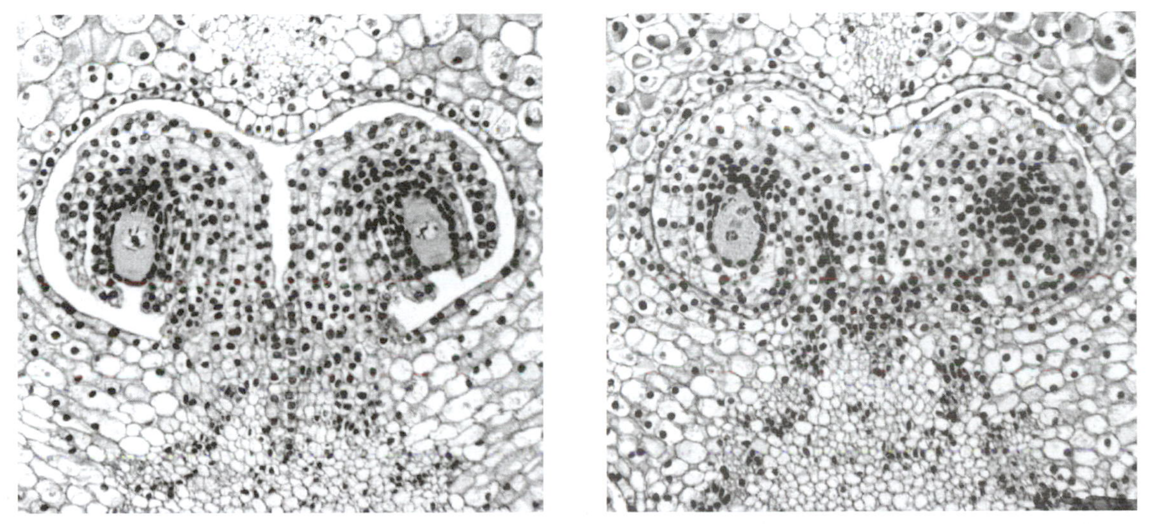

FIGURE 14.18 *Lilium* ovule, first meiotic division, X 90.

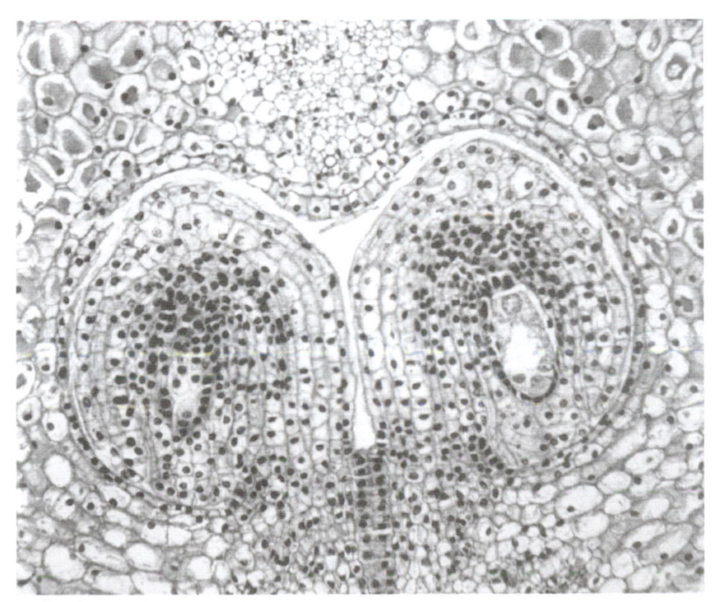

FIGURE 14.19 *Lilium* ovule, mature embryo sac, X 100.

Pollination and Fertilization

The culmination of the development of the gametophytes is the fertilization of the egg and the beginning of a new generation of plants. In the seed plants, this is preceded by pollination. Pollination in simply the transfer of pollen from the male reproductive structure to the female structure. In the flowering plants, this is achieved in a variety of ways. Most common is insect pollination, where honeybees, butterflies, flies, beetles, and many other insects get pollen on their body and carry it from one flower to another. Some flowers are very specialized as to the type of insect that is involved, i.e., the snapdragon from today's exercise or certain orchids. Other flowers may be pollinated by birds, bats, or other animals. Most flowers that are animal pollinated have bright colors and/or fragrant aromas that serve as beacons, attracting the pollinator. Other flowers may be very inconspicuous because they lack brightly colored petals. These are usually wind pollinated.

When a pollen grain lands on the stigma of a receptive pistil (**pollination**), it completes its development. Remember that there were two cells present? The **tube nucleus** is apparently involved in the growth of a slender tube that penetrates through the tissue of the style. The generative nucleus then divides once by mitosis, producing two **sperm nuclei**. This now represents a mature male gametophyte plant in the division Anthophyta.

Typically, the pollen tube somehow locates the micropyle of the ovule and then discharges the two sperm nuclei into the embryo sac. One of these fuses with the egg, producing the diploid **zygote**. The other sperm nucleus joins with the two polar nuclei, resulting in a triploid (3N) nucleus known as the **endosperm nucleus**. This process is known as "double fertilization."

The endosperm nucleus then divides rapidly, filling the embryo sac. The zygote becomes the **embryo** and, as it develops, relies on the endosperm as a food source.

Examine the prepared slides of pollen grains in various stages of germination. Try to find one that has a well-developed pollen tube with three nuclei in it. The tube nucleus will be the one near the tip of the tube, and the other two nuclei will be the sperm nuclei.

Make a labeled diagram of a germinated pollen tube with its **tube nucleus** and two **sperm nuclei**. (Fig. 14.20).

FIGURE 14.20 Pollen grain with pollen tube.

FLORAL EVOLUTION TALLY FORM

Each box should have either a dash (–) or a numeral in it.

A higher total indicates a more primitive flower structure.

		WATER LILY	GLADIOLUS	TULIP	GERANIUM	SNAPDRAGON	ROSE	DAISY RAY	DAISY DISK	DANDELION
Shape of Receptacle										
Extremely elongated	(5)	5								
Moderately elongated	(3)	–								
Flattened or concave	(1)	–								
Arrangement of Parts										
Obviously spiral	(5)	5								
Moderately spiral	(3)	–								
Whorled	(1)	–								
Number of Floral Parts										
Ten or more sepals	(2)	–								
Ten or more petals	(2)	2								
Ten or more stamens	(2)	2								
Ten or more carpels	(2)	2								
Less than ten sepals	(1)	1								
Less than ten petals	(1)	–								
Less than ten stamens	(1)	–								
Less than ten carpels	(1)	–								
Parts absent	(0)	–								
Fusion of Parts										
Free sepals	(2)	2								
Free petals	(2)	2								
Free stamens	(2)	2								
Free carpels	(2)	2								
Fused sepals	(1)	–								
Fused petals	(1)	–								
Fused stamens	(1)	–								
Fused carpels	(1)	–								
Parts absent	(0)	–								
Symmetry										
Radial symmetry	(5)	5								
Moderate bilateral symmetry	(3)	–								
Strong bilateral symmetry	(1)	–								
Ovary Position										
Hypogynous	(5)	5								
Perigynous	(3)	–								
Epigynous	(1)	–								
TOTALS		**35**								

Questions

1. How does seed production in the flowering plants differ from that of the conifers? (28)

2. What is the function of the following: (29)
   Sepals:

   Petals:

   Stigma:

   Ovary:

   C. Explain the relationship between a carpel and a pistil. (30)

4. Which is the most primitive flower of those examined in class? (31)

   Which is the most advanced?

5. How do ray flowers differ from disk flowers (three ways)? (32)

	Ray Flowers	Disk Flowers
a.		
b.		
c.		

6. How does the female gametophyte of a flowering plant differ from that of a conifer? (33)

7. In general, flowering plants produce less pollen than conifers. Why? (34)

8. Explain the difference between pollination and fertilization. (35)

Division Anthophyta: Flowering Plants — Fruits and Seeds

Introductory Notes

*After fertilization, changes occur in both the ovule and the ovary. The result of these changes is the development of the **seeds** and the **fruit**.*

SEEDS

Seeds are matured ovules. They contain an embryonic plant and a stored food in the form of **endosperm** tissue (in some the endosperm has been "used up" by the developing embryo, and large **cotyledons** are present instead). The integuments of the ovule have hardened to become a protective **seed coat** or **testa**.

The Garden Bean (*Phaseolus vulgaris*)

Obtain a bean seed that has been soaking in water. Compare the size of this seed to that of dry seeds. The difference is due to the imbibition of water by the starch in the cotylesons. This imbibition eventually causes the seed coat (**testa**) to rupture, and the primary root can then emerge and start the germination process by establishing the seedling in the soil.

GOALS

After completing this exercise, the student should be able to:

- give a concise definition of a seed.
- explain the difference between a seed with much endosperm and one with no endosperm.
- describe the structure of both a bean seed and a corn seed.
- identify the parts of a bean and corn seed.
- give a concise definition of a fruit.
- identify the fruit types listed in the key.
- differentiate between simple and compound fruits.
- list the main types of fleshy fruits, and explain how these types differ from each other.
- give a detailed description of the structure of an apple.
- answer the questions in the exercise.

Examine the soaked seed and locate the testa and hilum (a scar that indicates where the seed was attached to the placenta). With the aid of a dissecting microscope, the micropyle can be seen easily also. Figure 15.1 is an outline of a garden bean seed. Complete the diagram by drawing in and labeling the **hilum** and **micropyle**.

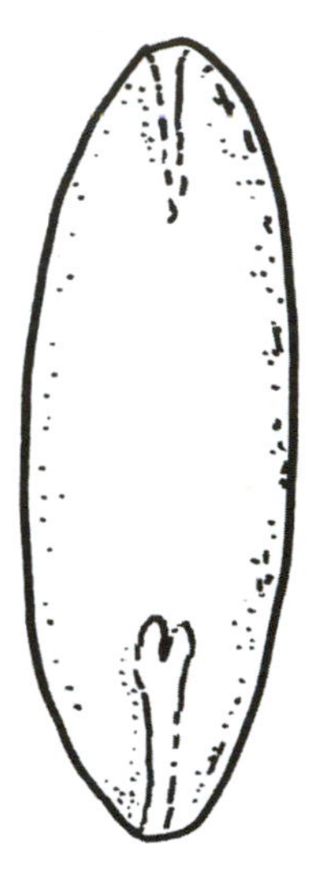

FIGURE 15.1 Exterior of garden bean seed, X 3.

Remove the seed coat, and separate the bean into two halves. The two large structures are **cotyledons**. Notice the well-developed embryo that is located between the two cotyledons. There is no endosperm present in the garden bean seed. What has happened to it?
(1)

Carefully remove the embryo, and examine it under a dissecting microscope. The large conical structure is the **hypocotyl**. The lower portion of the hypocotyl (toward the tip) is the **radicle** or embryonic root. The remaining or upper portion of the hypocotyl forms the lower part of the stem. The two small, folded leaves comprise the **plumule** or embryonic shoot. Notice that these leaves are already well developed, with conspicuous venation.

Figure 15.2 is a photograph of one-half of a bean seed. Label the **cotyledon, plumule, hypocotyl,** and **radicle**.

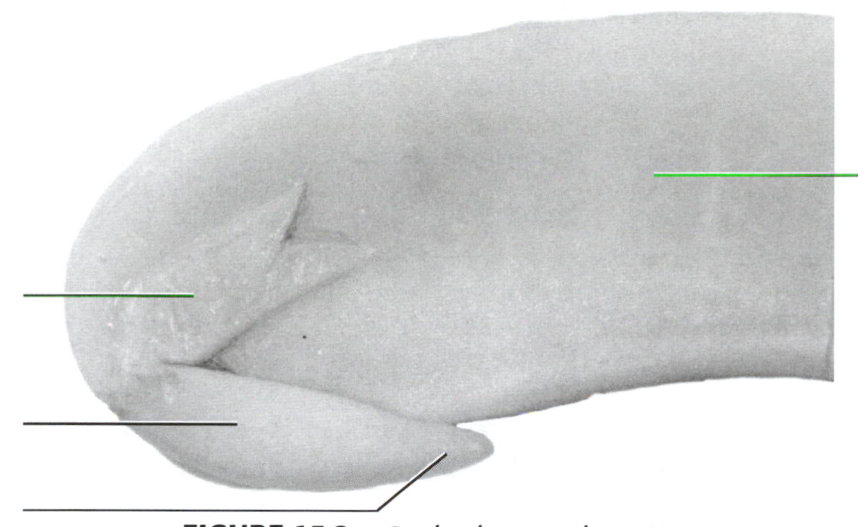

FIGURE 15.2 Garden bean embryo, X 6.

The Castor Bean (*Ricinus communis*)

Obtain a seed which has been soaked in water. Castor bean seeds have a strong resemblance to a swollen tick; in fact, *Ricinus* means tick in Latin. At one end of the seed is a swollen, somewhat spongy structure called the **caruncle.** This structure absorbs and retains water during germination of the seed. The **hilum** is a triangular or oval scar near the caruncle. The micropyle is in the same region but is too small to be seen. The line extending up one side of the seed is the **raphe,** formed by a fusion of the ovule stalk to the developing seed coat.

Label the **caruncle, hilum,** and **raphe** in the photograph of a castor bean seed (Fig. 15.3).

Crack and remove the brittle seed coat. When this is done, you should find a thin, papery, grey covering surrounding the inner contents of the seed. Remove this covering. The white, oily structure is the **endosperm.** What is its function? (2)

This endosperm is the source of castor oil, which is used not only as a purgative, but also as a high quality lubricant. The endosperm also contains an alkaloid called ricinin, which is a **very toxic** substance. Ingesting a single seed may be fatal to a child; two to eight for an adult.[1] The seeds have been soaked in several changes of water, but you should wash your hands thoroughly after this exercise.

Now, use your thumbnails to split the seed into two halves. Divide it in a plane parallel to the flattened surface. This should expose two very thin **cotyledons.**

[1] Schmutz, E. M., and L. B. Hamilton. 1988. *Plants that poison: An illustrated guide to plants poisonous to man.* Flagstaff, Arizona: Northland Press.

FIGURE 15.3 Castor bean seed, X 7.

Close examination will reveal three main veins on each. When the seed germinates, these cotyledons become large photosynthetic structures that remain on the plant for a long time. The bulk of the seed is the **endosperm** tissue.

At one end of the cotyledon is a small oval structure. This is the hypocotyl. The plumule is very poorly developed in the castor bean.

Compare the size of the cotyledons of the castor bean with those of the garden bean. (3)

Consider the size of the cotyledons, the stage of development of the embryo, and the amount of endosperm in the castor bean, and compare these to the garden bean. How will these affect the speed of germination of the seed? (4)

Label **endosperm**, **cotyledons**, and **hypocotyl** in the photograph of a castor bean seed which is split open to show the interior structures (Fig. 15.4).

Corn (*Zea mays*)

A kernel of corn is often erroneously referred to as a seed. It is really a fruit containing a single seed. Therefore, the terms *hilum* and *micropyle* cannot be used. The scar at the base of the kernel where it was attached to the ear is simply referred to as a fruit scar. The fruit wall, or **pericarp**, is thin, transparent, and fused with the seed coat.

Obtain some soaked corn grains. First, examine the external features. On one of the flat sides, you will see a white, shield-shaped region. This is where the embryo lies, inside the kernel. Most of this white area is the single cotyledon (which is usually referred to as the **scutellum**). In front of the scutellum, you should see a pointed ridge. The upper part of this ridge represents the **plumule**; the lower part is the **hypocotyl**. The rest of the kernel (the yellow portion) is composed of **endosperm** tissue.

Label these four regions in the diagram showing the surface view of a corn grain (Fig. 15.5).

To see the embryo in more detail, make a *longitudinal section* of the kernel through the center of the embryo. Make a second cut parallel to the first to provide a flat surface. Examine the cut surface under the dissecting microscope.

Add a drop of iodine solution (IKI) to the cut surface. Iodine reacts with starch, turning it a deep purple color. Where is the highest concentration of starch? (5)

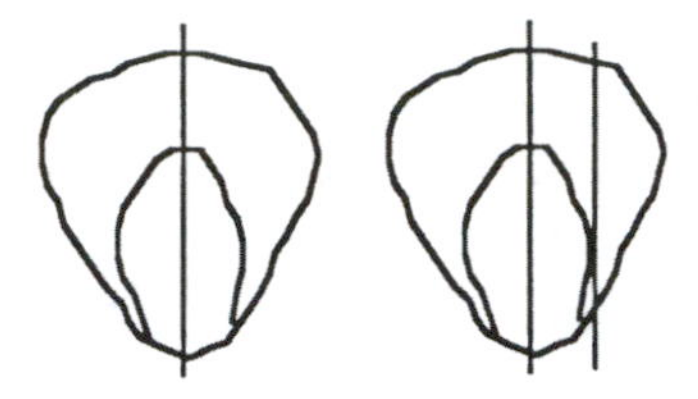

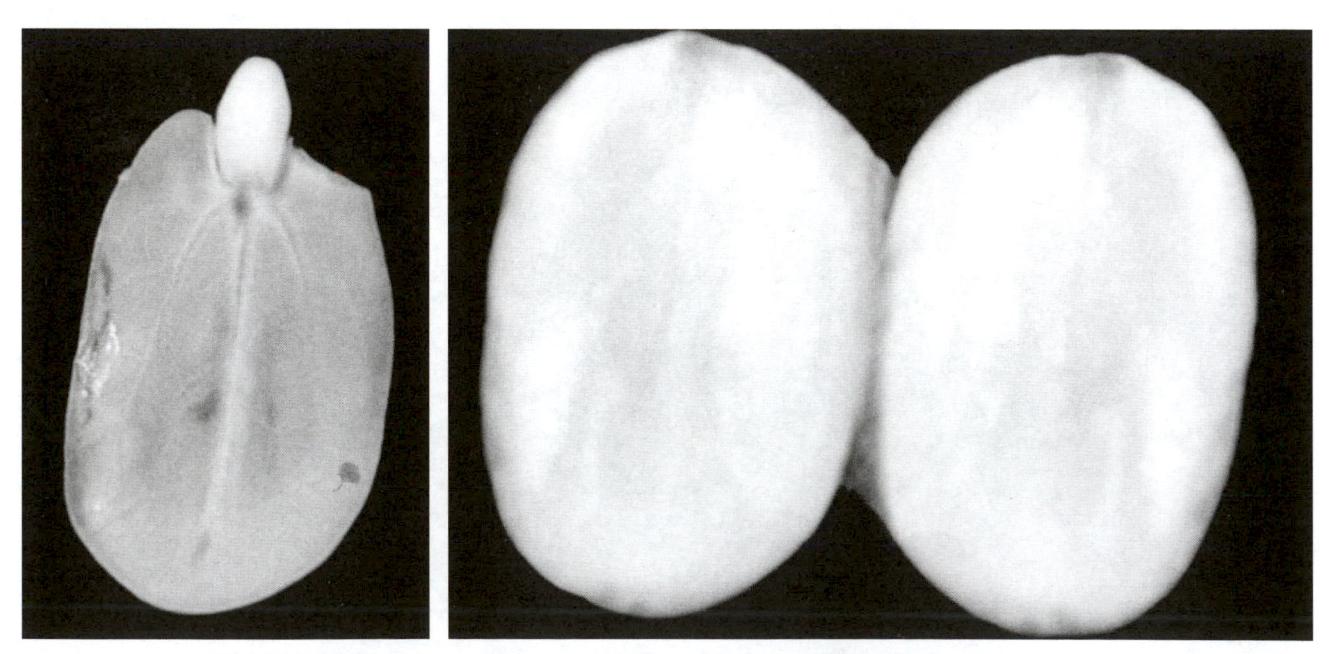

FIGURE 15.4 Castor bean seed opened to show the embryo, X 5.

The hypocotyl in this embryo is only a short, fat segment that connects all the other parts. The pointed structure extending upward from the hypocotyl is the **plumule**; the lower part of the hypocotyl is the **radicle.** A modified leaf, the **coleoptile,** protects the pumule, and the radicle is surrounded by a **coleorhiza.**

The corn cotyledon (**scutellum**) is a digestive body that never leaves the seed. It is attached to the hypocotyl and is adjacent to the endosperm tissue.

Figure 15.6 is a photograph of a longitudinal section of a corn kernel. Label the **radicle, plumule, coleoptile, coleorhiza, scutellum, pericarp,** and **endosperm.**

FRUITS

A fruit is a ripened or matured ovary with any other associated floral parts that may be fused to it. Fruits are found only in the Anthophyta. They may be variously modified, not only for protection of the seeds, but also for seed dispersal. Often the ovary wall undergoes profound changes, and three distinct regions can be recognized: an outer **exocarp**, a middle **mesocarp**, and an inner **endocarp**. All three comprise the **pericarp**.

There are more than a dozen different fruit types, and they may be divided into various categories. For example, one common way of categorizing fruits is by their texture (fleshy versus dry fruits). Another way is to separate them into "true" fruits, which are composed entirely of ovary tissue, and accessory fruits, which have floral parts in addition to the ovary. A third way, simple versus compound, is given in the first couplet of the following key.

Use the following key to identify the fruit types provided. Complete the list given at the end of the key.

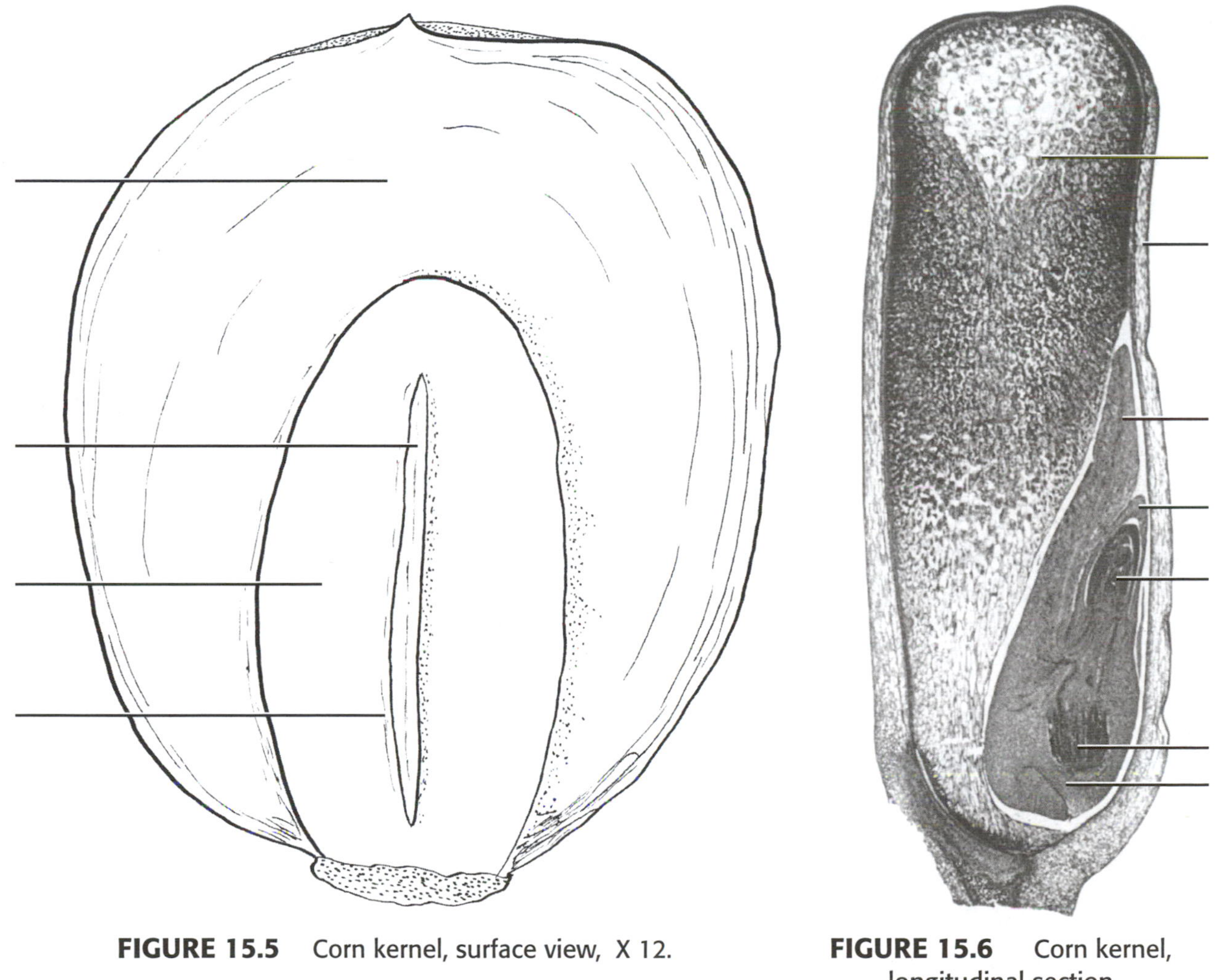

FIGURE 15.5 Corn kernel, surface view, X 12.

FIGURE 15.6 Corn kernel, longitudinal section through the embryo, X 12.

Courtesy Turtox/Cambosco Macmillan Science Co., Inc. Chicago, Illinois 60620

KEY TO VARIOUS FRUIT TYPES

1. Simple fruits — derived from a single ovary of a single flower ---------- 2
1. Compound fruits — derived from either the ovaries of several flowers or from several ovaries of a single flower ---------- 11

  2. Fruits obviously fleshy at maturity ---------- 3
  2. Fruit dry at maturity ---------- 5

3. Fruit wall inside the epidermis entirely fleshy, usually more than one seed ---------- Berry*
3. Fruit wall inside the epidermis not entirely fleshy ---------- 4

  4. Endocarp membranous, most of the fruit composed of a fleshy receptacle ---------- Pome
  4. Endocarp hard and stony making up a "pit," usually only one seed, not an accessory ---------- Drupe

5. Indehiscent fruits (not opening at maturity) ---------- 6
5. Dehiscent fruits (splitting open at maturity) ---------- 9

  6. Pericarp with winglike extension(s) ---------- Samara
  6. Pericarp lacking wings ---------- 7

7. Pericarp very hard and stony, one seed not fused to the pericarp ---------- Nut
7. Pericarp never stony (may be hard or papery) ---------- 8

  8. Seed fused to the pericarp ---------- Grain or Caryopsis
  8. Seed not fused to the pericarp (may be attached at one point) ---------- Achene

9. Composed of one carpel ---------- 10
9. Composed of more than one carpel, fused together ---------- Capsule

  10. Opening along one suture ---------- Follicle
  10. Opening along two sutures ---------- Legume

11. Fruit derived from several ovaries of several flowers, fused together ---------- Multiple Fruit
11. Fruit derived from several ovaries of a single flower ---------- 12

  12. Fruit with a prominent fleshy receptacle ---------- Aggregate-accessory fruit
  12. Fruit lacking a fleshy receptacle ---------- Aggregate fruit

* There are several subcategories of berries, such as the **Hesperidium**, characteristic of citrus fruits, and the **Pepo** is typical of members of the squash family.

Plant	Fruit Type	Plant	Fruit Type	Plant	Fruit Type
Peanut	Legume	Pineapple		Almond	
Pecan		Iris		Acorn	
Apple		Corn		Tomato	
Strawberry		Walnut		Milkweed	
Maple / Ash		Coconut		Bl. Locust	

Tomato (*Lycopersicon esculentum*)

1. Look carefully at the end away from where the tomato was attached to the stem. You should be able to see a very small prickle present. This is the remains of the stigma and style.
2. Is the tomato flower epigynous or hypogynous? (6)
3. To examine the interior of the tomato, cut one in half, crossways. Notice that this fruit has only a slightly toughened skin and a completely fleshy interior.

   What type of fruit is this? (7)
4. Make a diagram of a cross section of a tomato, using the outline of the circle to represent the exocarp. Label the **placenta**, **seeds**, **locule**, **exocarp**, and the combined **mesocarp** and **endocarp** (Fig. 15.7).

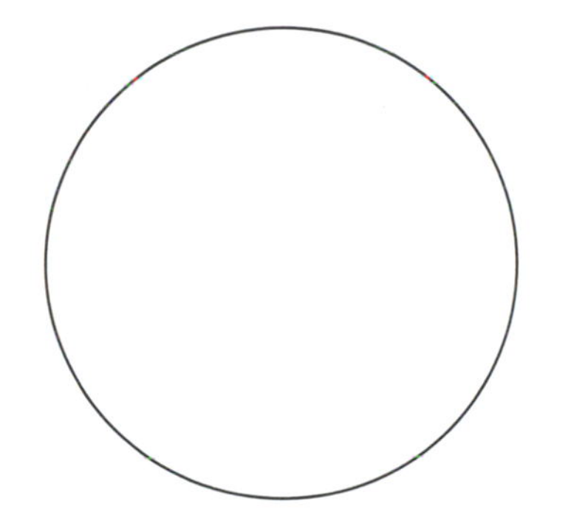

FIGURE 15.7 Tomato, cross section.

Sunflower (*Helianthus* sp.)

1. Examine a sunflower "seed." Remember the structure of a daisy flower, which was an epigynous flower. Notice the scars on each end. The more narrow region is where the ovary was attached to the receptacle, and the more rounded end is where the floral parts were attached. Carefully split open the "seed." (By now you should have concluded that this structure is actually a fruit.) Notice that the true seed is attached to the fruit wall at only one place — at the bottom of the fruit. Leave the true seed intact and in place in one-half of the fruit.
2. Make a diagram of one-half of a single sunflower fruit, showing the true seed. Label the **seed** and the **pericarp**. (Fig. 15.8).

FIGURE 15.8 Sunflower fruit with seed.

3. What type of fruit does the sunflower produce? (8)

Apple (*Malus* sp.)

1. Use the dissecting microscope to examine the blossom end of the apple. Notice the remains of the sepals and stamens at this end. Is the apple flower hypogynous or epigynous? (9)
2. Working with a partner, use one apple and make a longitudinal section of the fruit, being certain to go through the center of the end away from the stem.
3. The shiny, membranous core surrounding the seeds (which are in **locules**) is the **endocarp**. The thin green line about 1/2 inch from the endocarp is vascular tissue that went around the ovary, up to the sepals, petals, and stamens. This line marks the approximate location of the **exocarp**. The region between these vascular bundles and the endocarp is the **mesocarp**.
3. Most of the apple is derived from tissue other than the ovary.
   a. What floral structure develops into the edible part of the apple? (10)

   b. What type of fruit is the apple? (11)
4. Label the **endocarp**, **exocarp**, **locule**, **mesocarp**, **receptacle**, **seed**, and **sepals** in the diagram of a longitudinal section of an apple (Fig. 15.9).

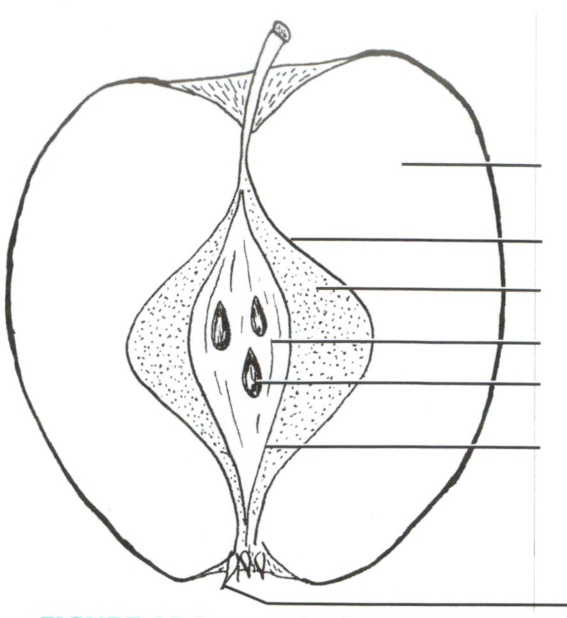

FIGURE 15.9 Longitudinal section of an apple.

5. Make a cross section of the fruit.

   a. How many carpels were present in the pistil? (12)

   b. Is the apple a dicot or a monocot? (13)

6. Locate the various structures mentioned previously.
7. Make a diagram of a cross section of an apple, showing and labeling the **receptacle, exocarp, mesocarp, endocarp, seed, sepals,** and **locule** (Fig. 15.10).

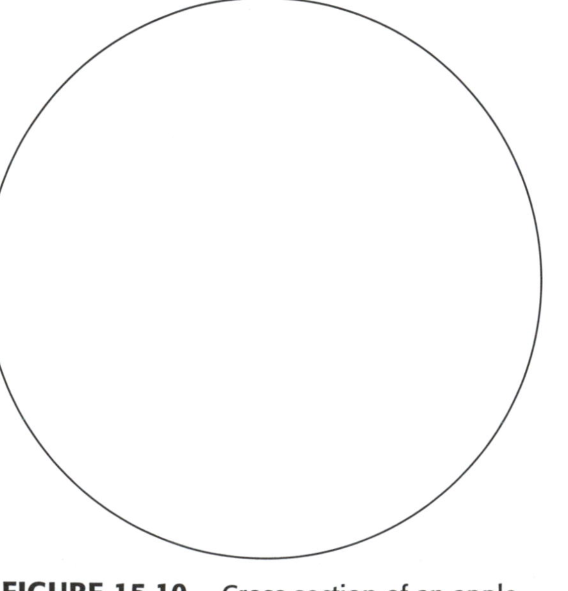

FIGURE 15.10 Cross section of an apple.

Strawberry (*Fragaria* sp.)

Your strawberry probably has sepals attached. This should tell you whether the flower was hypogynous or epigynous. Which of these two is the case? (14)

1. Remove the calyx, and examine the region which was next to the fruit. Notice the numerous stamens that are still present.
2. Make a longitudinal section through the center of the fruit, and examine the cut surface. The large, fleshy part is the enlarged receptacle. Notice the pink lines extending from the receptacle to the exterior of the fruit. These are vascular bundles. To what structures are these vascular bundles headed?
3. If you answered "seeds," you made a common error. Those small structures on the surface of the fruit are not seeds, but are the mature ovaries of the flower. Each of these contains a single seed. Examine these with the dissecting microscope, and you will see the enlarged ovary, the stigma, and the style of the pistil. These "seeds" are achenes, a simple fruit that develops from a single carpel.

   What type of fruit does the strawberry have? (15)

4. Make a diagram of the cut surface of the strawberry, labeling the **receptacle**, a **vascular bundle,** and an **achene** (one of the "seeds") (Fig. 15.11).

FIGURE 15.11 Longitudinal section of a strawberry.

Other fruits are available, either as demonstration specimens or as additional examination and discussion specimens. Follow the directions of your instructor for these.

Questions

1. Is the iris a monocot or a dicot? (16)

   How can you tell by examining the fruit?

2. Corn kernels, peanuts, sunflower seeds, strawberries, almonds, and coconuts are all often misnamed. Why? Explain what the problem is for each one (and what the correct explanation is). (17)

	Misconception	Botanical Truth
Corn kernel		
Peanut		
Sunflower Seed		
Strawberry		
Almond		
Coconut		

3. What do milkweed seeds and dandelion seeds have in common? How are they different? (18)

   Common feature:

   Differences:

   Dandelion:

   Milkweed:

4. Which seed has the best developed embryo? What correlation is there between the amount of endosperm and the development of the embryo? (19)

   a.

   b.

5. What are two possible functions of fruits? (20)

   a.

   b.

6. What plant structure develops from the plumule? (21)

In dicots, what two structures develop from the hypocotyl?

7. How does the function of the corn cotyledon differ from the function of the bean cotyledon? (22)

Corn:

Bean:

8. What are some important evolutionary advances that have occurred in the various plant groups? (23)

Algae:

Bryophytes:

Primitive land plants:

Ferns:

Gymnosperms:

Angiosperms:

APPENDIX

A

Two Classification Schemes

	Tippo (1942)			Bold (1957)
Subkingdom	Thallophyta			*(abandoned)*
Phylum 1	Cyanophyta	Blue-green Algae	Division 1	Cyanophyta*
Phylum 2	Chlorophyta			
Class	Chlorophyceae	Green Algae	Division 2	Chlorophyta
Class	Charophyceae	Stoneworts	Division 3	Charophyta
Phylum 3	Euglenophyta	Euglenoids	Division 4	Euglenophyta
Phylum 4	Phaeophyta	Brown Algae	Division 5	Phaeophyta
Phylum 5	Rhorophyta	Red Algae	Division 6	Rhodophyta
Phylum 6	Chrysophyta		Division 7	Chrysophyta **
Class	Xanthophyceae	Yellow-green Algae	Class	Xanthophyceae
Class	Bacillariophyceae	Diatoms	Class	Bacillariophyceae
Class	Chrysophyceae	Golden-brown Algae	Class	Chrysophyceae
Phylum 7	Pyrrophyta	Dinoflagellates	Division 8	Pyrrophyta
Phylum 8	Schizomycophyta	Bacteria	Division 9	Schizomycota
Phylum 9	Myxomycophyta	Slime Molds	Division 10	Myxomycophyta
Phylum 10	Eumucophyta	True Fungi		(abandoned)
Class	Phycomycetes	Algalike Fungi	Division 11	Phycomycota***
Class	Ascomycetes	Sac Fungi	Division 12	Ascomycota
Class	Basidiomycetes	Club Fungi	Division 13	Basidiomycota
	Fungi Imperfecti	Imperfect Fungi		Fungi Imperfecti
		Lichens		
Subkingdom	Embryophyta			(abandoned)
Phylum 11	Bryophyta		Division 14	Hepatophyta
Class	Hepaticae	Liverworts	Class	Hepatopsida
Class	Anthocerotae	Horned Liverworts	Class	Anthoceropsida
Class	Musci	Mosses	Division 15	Bryophyta
Phylum 12	Tracheophyta			(abandoned)
Subphylum	Psilopsida	Psilophytes	Division 16	Psilophyta
Subphylum	Lycopsida	Club Mosses	Division 17	Microphyllophyta
Subphylum	Sphenopsida	Horsetails	Division 18	Arthrophyta
Subphylum	Pteropsida			(abandoned)
Class	Filicinae	Ferns	Division 19	Pterophyta
Class	Gymnospermae			(abandoned)
Subclass	Cycadophytae	Cycads	Division 20	Cycadophyta
Subclass	Coniferophyta			(abandoned)
Order	Ginkgoales	Ginkgo	Division 21	Ginkgophyta
Order	Coniferales	Conifers	Division 22	Coniferophyta
Order	Gnetales	Gnetum, etc.	Division 23	Gnetophyta
Class	Angiospermae	Flowering Plants	Division 24	Anthophyta

* As indicated in Exercise 7, these organisms are now categorized as Cyanobacteria.

** The three classes of the Chrysophyta are often elevated to division status. This is the approach taken in this manual, with *Vaucheria* placed in the division Xanthophyta and the diatoms placed in the division Bacillariophyta.

*** The division Phycomycota is rarely used now. The three genera studied as representatives of the group are more often separated into three different divisions — in two kingdoms.